Measurement Matters

Also Available from ASQ Quality Press:

Performance Measurement Explained: Designing and Implementing Your State-of-the-Art System
Bjørn Andersen and Tom Fagerhaug

Work Overload: Redesigning Jobs to Minimize Stress and Burnout
Frank M. Gryna

Finding the Leader in You: A Practical Guide to Expanding Your Leadership Skills
Anton G. Camarota

Linking Customer and Employee Satisfaction to the Bottom Line
Derek Allen and Morris Wilburn

The Trust Imperative: Performance Improvement Through Productive Relationships
Stephen Hacker and Marsha Willard

Making Change Work: Practical Tools for Overcoming Human Resistance to Change
Brien Palmer

Improving Performance Through Statistical Thinking
ASQ Statistics Division

Office Kaizen: Transforming Office Operations into a Strategic Competitive Advantage
William Lareau

The Change Agent's Guide to Radical Improvement
Ken Miller

Measurement Matters

How Effective Assessment Drives Business and Safety Performance

Brooks Carder and Patrick Ragan

ASQ Quality Press
Milwaukee, Wisconsin

American Society for Quality, Quality Press, Milwaukee 53203

Printed in the United States of America

12 11 10 09 08 07 06 05 04   5 4 3 2 1

Library of Congress Cataloging-in-Publication Data

Carder, Brooks, 1942-
Measurement matters : how effective assessment drives business and safety performance / by Brooks Carder and Patrick Ragan.
p. cm.
Includes bibliographical references and index.
ISBN 0-87389-634-3 (hard : alk. paper)
1. Industrial management--Statistical methods. 2. Industrial safety--Statistical methods. 3. Deming, W. Edwards (William Edwards), 1900- I. Ragan, Patrick, 1954- II. Title.

HD30.215.C365 2004
363.11'02--dc22

2004015354

ISBN 0-87389-634-3

Publisher: William A. Tony
Acquisitions Editor: Annemieke Hytinen
Project Editor: Paul O'Mara
Production Administrator: Randall Benson
Special Marketing Representative: David Luth

Printed on acid-free paper

ASQ
AMERICAN SOCIETY FOR QUALITY™

Quality Press
600 N. Plankinton Avenue
Milwaukee, Wisconsin 53203
Call toll free 800-248-1946
Fax 414-272-1734
www.asq.org
http://qualitypress.asq.org
http://standardsgroup.asq.org
E-mail: authors@asq.org

Table of Contents

Preface

We believe few, if any, significant efforts in business are completed by just one person. This book grows from that belief. It is truly the intertwined effort and experience of both of us that contributes to this book. Our histories, training, and ideas came together to create what we hope you find is both a scientific and practical discussion of how to measure, understand, and improve some very complex processes in business and in life. We add life here because of a comment from one of our editors, Marlene Bergeron:

> *I actually have taken what we have learned and have been able to apply it to my personal life. One area this really helps is during arguments and misunderstandings. I can stick to the "root cause" of the disagreement, so to speak. Instead of falling into old baggage discussions, . . . I can stay focused on the facts and avoid coming to the wrong conclusions. You might even say it is something similar to a personal-type version of the causal tree.*
>
> *I bring this up, because this is how powerful a tool the book can be. Not only can people learn how to create a better workplace, they can use the same techniques in their personal lives.*

We owe a debt of gratitude to many people for contributing their ideas and challenging our ideas. They helped us clarify our ideas and make them more meaningful. Of course our greatest debt is to the late Dr. W. Edwards Deming, who was an inspiration to both of us. He strongly encouraged Brooks to write about quality from the perspective of a social scientist.

Marilyn Monda, a mathematical statistician and Six Sigma Master Black Belt, has worked with us off and on for many years and provided a technical review of several chapters. Marlene Bergeron provided invaluable editing of the entire work.

We have used the vehicle of measuring and improving safety to explicate our ideas. Safety management, or accident prevention, is a complex, difficult-to-measure process. We have spent a lot of time and effort working on this. We believe the knowledge and tools provided in this book can be applied to many other complex systems in business.

At the time of this writing, the two of us have been learning and working together for over 12 years. Our relationship began through the lens of measurement. Pat had assembled a team of managers and executives to measure and improve safety performance. Because the company had many small plants, the team understood that incident rates would be a poor measure. The committee hired Brooks to present a seminar on safety measurement. This seminar led to a pilot project and eventually to many years of our working together to improve the safety-management process. Although our individual careers have taken many turns since then, our relationship and our colleagueship have endured.

The two of us have followed very different career paths. Brooks was trained as an experimental psychologist. His 40-year career includes academia, consulting, and overall management of a small (120 employees) business. Pat was trained as a safety engineer and later obtained an MBA. His 25-year career has been one of service in the front lines of corporate safety management. We believe that our life experiences are complimentary and have made us much more productive as a team. At the same time, this has created some awkwardness in writing the book. Many of the experiences that we bring to this are individual ones. This has caused us to sometimes use the first person singular, and we have not found a way around some clumsiness in style. We hope the reader will forgive us.

This book is a presentation of what we have learned in our work. We believe that the lessons are significant and can help managers do better, no only in safety but in every aspect of management and leadership. Our collaboration began with a focus on measurement, and that is carried throughout this book. We have learned over the years that what we measure and how we measure it drives our actions, for better or worse. It is often said, "What gets measured gets done." Unfortunately, the truth of this often creates serious problems. Many people are measuring the wrong things and, based on their measurements, taking the wrong actions. If we measure the right thing in the right way, we gain insight and leverage for positive, lasting change. If we measure the wrong things or measure in the wrong way, only luck will help us. More likely we will waste time and resources. In safety, these measurement mistakes leave people at risk.

Much of what is measured in business, especially in safety, is governed by conventional wisdom. In many cases, there are absolutely no data to support the conventional wisdom, and often the evidence that does exist indicates that the conventional wisdom is wrong. We hope this book will improve on the conventional wisdom and help readers avoid misdirected decisions driven by flawed measures.

One of the many things we learned from Dr. Deming was that there is no knowledge without theory. The purpose of our book is to help others make better decisions because they have good measures based on sound theory of effective measurement. Rather than trying to lay out a set of procedures or practices that can used to improve safety, we hope to provide the theoretical knowledge that will enable the reader to understand and address the underlying causes of management-system problems. The stories we tell are illustrations and are not intended as recipes for the reader to follow. The reader must consider the application of the theory to each particular situation. The book discusses a number of measurement processes that are employed in safety, and uses the theory of measurement to evaluate the strengths, weaknesses, and appropriate application of each.

We hope that some of what we write will become the new conventional wisdom. We also look forward to the inevitable improvement of knowledge in this field that will reveal *our* limitations and advance everyone to that next level of understanding and capability.

Brooks Carder
Pat Ragan
March 2004

1

The Fundamental Principles of Dr. Deming's Approach

Many years ago, in my first year of graduate school, I was in a seminar on abnormal psychology. The topic turned to hysterical blindness, a condition in which a patient with an apparently intact visual system is nevertheless unable to see. One of the students offered a simple cure. He would put such patients in a situation in which they would have to use visual cues to avoid painful electric shock. He was certain that a miracle cure would ensue. He was accused of being a right-wing psychotherapist. He later became a world-famous clinical psychologist, but so far as we know he never tried this treatment. However, he was suggesting a management principle that is still in vogue: If they won't do it, we will make them do it.

Sometimes extreme pressure is required to produce positive change. More often, extreme pressure fails and frequently causes damage to individuals and organizations.

This book is about is how to create positive change in organizations and individuals. We often see hysterical blindness in both. What is wrong is obvious to us, but they may not be able to see it. Our challenge is to find leverage for change. Leverage implies a way to use our knowledge and influence to greatest advantage in creating positive change. Usually when we are trying to change people and organizations, applying strong force generates an equal force in the opposite direction, consuming energy and resources and usually preventing change. As an example, we have seen this nonproductive consumption of energy and resources many times in conflicts between unions and management. Each is reacting to the pressure applied by the other.

Such systems are very resistant to change. In the union-versus-management scenario, the situation typically does not change until management, which is typically the party with the most direct leverage, decides to change it and takes steps to confront the conflict or compromise to resolve it. A former senior manager at Ford told us that the single aspect of Ford's quality transformation in the 1980s that gave him the most pride was the dramatic change in management's relationship with the union. He said it was typical to see executives and hourly workers seated side by side in a classroom, something that would have been unheard of prior to the transformation.

The chairman of Ford during the early stages of the transformation was Donald Petersen. In a memoir of that period (Petersen and Hillkirk 1991), Petersen mentions that he frequently toured Ford plants. His critical criterion for evaluation was whether the plant manager knew the hourly workers. He said it was clear that some had memorized the names just before his arrival. Managers who did not know the workers did not stay with the company.

After graduate school, I spent several years as a scientist in a university. I then entered business and spent 13 years as a manager. Initially I tried the "my way or the highway" approach to management, but it did not work for me. I began to develop more cooperative relationships with my staff, and things went much better. However, the science of management was quite puzzling to me. Much of what was accepted as good management practice, including the use of strong threats as motivators, was contrary to the accepted scientific principles I had been taught and had taught to others.

In 1988, I encountered the work of Dr. W. Edwards Deming, the man who created a revolution in quality management in America. I had an epiphany. Dr. Deming's teachings made sense of many of the questions and confusions I had encountered in trying to reconcile what I knew as a scientist with what I saw as common practice in business. From 1988 to Dr. Deming's death in 1993, I had the opportunity to attend five of his four-day seminars and form a casual acquaintance with him. By that time Dr. Deming was famous as the man responsible for the rise of Japanese industry following World War II, but who had gone unheeded in America. Only after we had lost our industrial leadership to the Japanese was his knowledge recognized in America. In 1982 he was engaged by Ford. The programs that grew from his teachings helped move Ford from the brink of bankruptcy to far and away the most profitable U.S. auto manufacturer by 1995.

Many have suggested that quality programs were a flavor of the month that is now stale. That is probably because many companies treated them as a flavor of the month and failed to embrace the principles seriously and go through the discipline needed to establish a successful quality program. The companies that did embrace quality reaped substantial benefits. Easton and Jerrell (1998) conducted an in-depth study of quality programs and their results in American industry. Their findings indicate that "performance measured by both accounting variables and stock returns is improved for firms adopting TQM. The improvement is consistently stronger for firms with more advanced TQM systems." Dr. Deming's ideas are at the heart of lean manufacturing[1] and Six Sigma,[2] both very successful current approaches to the improvement of business performance.

Our experience with Dr. Deming and his teachings led us to adopt the principles and methods he taught. The Deming approach depends on some fundamental shifts in thinking, away from what is still common and traditional in management. Dr. Deming's assumptions are, of course, much better supported by scientific evidence than are many practices that were and still are popular in business. However, the impact of science on common belief and practice is often very late in coming. Galileo was not the first and certainly not the last scientist to experience a rather pointed rejection of what seemed to him (and, with the benefit of hindsight, to us) obvious scientific truth. Usually change does not happen until it is demanded by the marketplace. Einstein's theories were widely disputed until the power of nuclear physics converted everyone. Few listened to Dr. Deming until the quality of Japanese products overwhelmed American products in the marketplace.

Dr. Deming's approach to management is founded on two fundamental principles. Although based in science, these principles form the philosophical framework of his system.

> ***Principle 1:*** *People will do their best to do what they think is expected of them.*

MUCH OF WHAT HAS EVOLVED AS THE PRACTICE OF MANAGEMENT AND SUPERVISION SEEMS TO DERIVE FROM THE ASSUMPTION THAT, GIVEN THE CHANCE, WORKERS WILL ATTEMPT TO AVOID DOING WHAT NEEDS TO BE DONE.

A fundamental responsibility of management is to ensure that the employees of the organization know what needs to be done in order to make the organization successful and that they do it. However, much of what has evolved as the practice of management and supervision seems to derive from the assumption that, given the chance, workers will attempt to avoid doing what needs to be done and will do something else altogether. Unfortunately, this assumption is often a self-fulfilling prophecy. Acting on this assumption, management institutes extensive control systems. Employees' work is checked, measured, audited, and rechecked. This is expensive and slows down the work. More important, it sends the message to employees that they are not trusted to do a good job. To the extent that management exerts control over what employees do, employees' pride in doing good work is diminished. Once this polarization between managers and employees is established—that the job of managers is to make certain that otherwise irresponsible employees do the right thing—it may be increasingly necessary for managers to do just that. We have had a great deal of personal experience with this phenomenon.

John Whitney, a professor at Columbia University and an associate of Deming, has written eloquently on this in *Trust Factor* (Whitney 1994). Whitney often used a simple question to ferret out excessive controls imposed by management: "If everyone here knew how to do his/her job properly and was trusted to perform it, what are we doing now that we could stop doing?"

When I first encountered Whitney's work in 1989, an associate and I had just taken over management of a company that sold promotional products. The company processed about 10,000 orders per year. Every order was different. There were about 1,000 different products sold, and each product had a corporate logo and some message that a particular customer wanted to promote. The company did not inventory the products. Rather, it ordered them from a set of about 50 major suppliers and over 500 minor suppliers. The suppliers imprinted the products with the logo and the message, and shipped to the customer. If a product, such as a cup, a pen, or a t-shirt, was imprinted improperly, it had virtually no salvage value. If it reached the customer too late for an event or a meeting, it was also virtually worthless. Thus, in spite of the complexity, accuracy and timing were very important. Therefore, we had many checks and double checks to make certain that everything was done properly. In 1988 the company had a fatal-error rate of about 2.5 percent of orders. These were errors that led to crediting the customer or redoing the product. Our suppliers made such errors at a rate of about 0.5 percent. The median order took five days from reaching the office to sending a purchase order to the supplier. Because many customers needed their goods right away, this created rush orders that got special handling. This, of course, added cost and further slowed down the regular orders. In 1988, 50 percent of the orders were rush orders.

I sat down with a small team of workers in the office and presented them with Whitney's question. They immediately provided a list of seven things they were doing that could stop if people were trusted to do their jobs properly. A second, independent team quickly gave me the same list. Most activities listed involved employees checking the work of other employees. We quickly eliminated these checks. At the time, some middle managers resisted the change. They argued that the company was taking great risk and that it would surely lose a lot of money from the increase in errors that was certain to result.

An obvious result of the changes was that orders were processed more rapidly. This made sense, as they had eliminated a number of steps in the process. The median processing time went from five days to one day. More important was the fact that errors declined as well. Errors made in the office went from 2.5 percent of orders to less than 0.5 percent. Instead of losing accuracy, eliminating checks actually increased it dramatically. Thus the office became more productive and more accurate. In 1988 they required 1.8 administrative employees to process $1,000,000 in orders. By 1995 it required 1.2 employees, who did it quicker and with far better accuracy.

Another result was dramatically improved morale. People began to enjoy work that was previously seen as drudgery. Between 1988 and 1995, only two persons out of our total staff of 30 left to work for another company. This amounts to a turnover of less than 2 percent annually. Years later employees remembered the process with vivid, positive feelings about it.

Of course, you might say that trusting employees is OK when we are talking about the wrong logo on a cup, but it will not wash when we are talking about the lives of thousands of people as in the manufacturing of airplane parts. Our response would be that much of what people do in the name of ensuring safety actually increases the probability that accidents will happen, just as much of the checking that we formerly did to ensure accuracy actually increased the likelihood of mistakes. Dr. Deming often pointed out that the Federal Aviation Administration (FAA) guaranteed that some aircraft accidents would happen by requiring two persons from the cockpit to report the position of commercial aircraft. This meant that responsibility for precise reporting was divided. Divided responsibility means that no one is responsible. Having one person check or approve another's work is a case of divided responsibility and will degrade accuracy rather than promote it.

If a check must be made, it must be completely independent. Rather than review a calculation, the calculation must be repeated independently and then compared. If we wanted to really improve the accuracy of orders through checking, we would have two people enter them independently and then compare them. If identical, they would go to the supplier for processing. If they differed, we would find and correct the problem. In the case of promotional products, this is too expensive for the additional accuracy we would achieve. We acknowledge that there are many circumstances in industry where this is necessary. However, in practice, we rarely see inspection carried out in this fashion. What we see is checking, reviewing the calculation made by another worker. This is the worst thing that could be done; it will make accuracy worse, not better, and will cost time that could be better spent in other ways.

ONE OF THE CRUCIAL ASPECTS OF THIS FIRST PRINCIPLE IS THAT PEOPLE WILL DO THEIR BEST TO DO WHAT THEY *THINK* IS EXPECTED OF THEM, NOT WHAT IS *ACTUALLY* EXPECTED OF THEM.

A crucial aspect of this first principle is that people will do their best to do what they *think* is expected of them, not what is *actually* expected of them. Having someone check others' work is a subtle way of telling those being checked that they are not trusted to do the job properly. At the very least, the process decreases each employee's responsibility for the outcome. In our experience, it damages morale as well. It is demoralizing to be assigned a task, do all the work required, but not be trusted to complete the job.

Another problem is created by incentive programs, merit review programs, and other such attempts to motivate and control employees. These programs often create expectations not intended by management. A company is a very complex system. It is not possible to write a set of rules and procedures that will properly guide all decisions and actions of employees and that will be in the company's best interests. For example, if an incentive plan specifies no accidents, that can be achieved by not reporting any even if they do occur. If it specifies a sales quota, then employees will try to meet it, even if they have to cut prices and eliminate profit altogether, or if they have to sell customers more goods than the customers can use, creating unhappy customers. For many years the pharmaceutical industry had a problem with growth targets. Because the industry was growing rapidly, companies usually met targets before the end of the year. Rather than exceed the target, managers held orders for entry after January 1. The procurement manager of one of the largest drug chains told us that as a result, no wholesaler in Chicago had stock of a particular key antibiotic in late December.

Nordstrom, the large clothing retailer, is legendary for the quality of its customer service. As a sometime customer, I can testify to the pleasant experience of shopping there and the ease with which problems are resolved. I once bought a blazer there, and over a year later the fabric of the lapel began to separate from the stiffening material underneath. When I went to complain, I was immediately told, "Whatever you paid for it can be used as credit on another one." Nordstrom employees are issued a card with Nordstrom's policy written on it: "Use your own best judgment at all times." Employees are expected to do what is best for Nordstrom and their customer, and they have the freedom to do it.

Many of us know from personal experience that having final authority changes our outlook. Years ago the company I worked for gave me the authority to authorize purchases well into six figures. If I turned a purchase order down, however, the decision could be appealed to a superior. Later, that superior was removed. When I was the final authority, I had to be much more thoughtful about the process. I recall signing several purchase orders that I would have previously rejected. Some might say that having final authority changed me from an annoying bureaucrat into a thoughtful executive. Certainly I believe that having final authority changed my thinking process for the better. In fact, bureaucracies are created by expanding rules and reducing authority. When the rule makes no sense, it must still be enforced, circumvented, or ignored. Enforcement of a nonsense rule destroys the credibility of other rules and encourages circumvention and ignoring of rules in general, even those that are there for very good reasons. That is the essence of bureaucracy. Although assigning authority without proper training and support can be misguided, most companies in our experience err on the side of giving too little authority and having too many controls.

Principle 2: *To improve performance you need to improve the system rather than focus on the individuals.*

DR. DEMING ARGUED THAT THE SINGLE MOST IMPORTANT THING HE TAUGHT IN JAPAN IN 1950 WAS THAT A BUSINESS SHOULD BE VIEWED AS A SYSTEM.

Dr. Deming argued that the single most important thing he taught in Japan in 1950 was that a business should be viewed as a system. The system included not only the various departments of the company but also external agencies, including suppliers and customers. In modern business we would include regulatory agencies, such as the Occupational Safety and Health Administration (OSHA), Environmental Protection Agency (EPA), and Securities and Exchange Commission (SEC), along with banks and other financial institutions. These parts must work together efficiently for the business to prosper. The diagram in Figure 1.1 is similar to the one Dr. Deming presented to the Japanese in 1950. He called this "the spark that ignited Japan."

The chief point made by the diagram is that the diverse parts of the company, including many that are not *in* the company, such as customers, form an interconnected system. You cannot act in one part of the system without influencing the others.

Dr. Deming argued that at least 85 percent of the variation in performance of a business derived from the system and at most 15 percent from the individuals in the system. When the majority of workers are performing poorly, the problem is in the system not the workers.

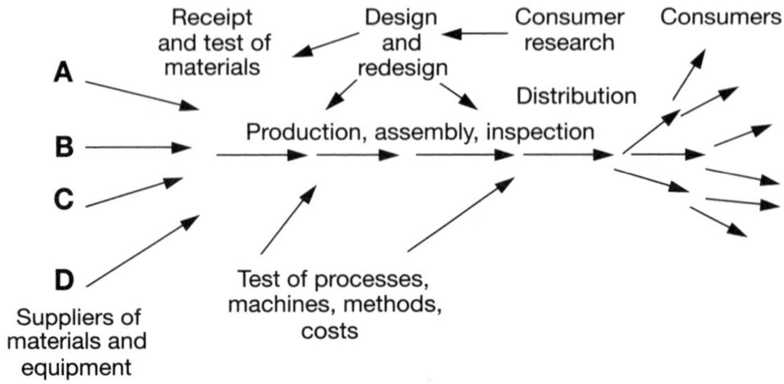

Figure 1.1 A business as a system.
Source: Deming, W. E., *The New Economics for Industry, Government, Education*. MIT Center for Advanced Engineering Study. Cambridge, MA, 1993. Used with permission.

It seems to be an American tradition to focus on individual performance. We are overwhelmed by the attraction of celebrity. We explain the performance of great teams by pointing to the coaches and the star players. We explain the entertainment value and money-making power of motion pictures by citing the stars who play in them. We explain the success of companies by pointing to the CEO. History would tell us that great teams create stars as often as stars create great teams, and that great movies create movie stars as often as stars create great movies. CEOs without a superior system cannot build a great company no matter how capable they might be as individuals.

Jim Collins, a consultant, wrote an important book with Jerry Porras called *Built to Last* (Collins and Porras 1994), a study of the characteristics of outstanding companies. Following its publication, some readers complained, "It's good to know what a great company is like, but our company is not great. Your book does not tell us anything about how to get there." In response, Collins conducted a five-year study of companies that had transformed themselves from mediocre to outstanding. The study was published in Collins's book *Good to Great* (Collins 2001). The criterion for inclusion in the study was that the company had to perform at or below the level of its competition for 10 years and then exceed the average performance of companies in its industry by a factor of three for 15 years. Only 11 public companies met this criterion. Collins and his research team then studied these companies to find common characteristics that could account for their phenomenal success.

Several common threads emerged. First, you have likely never heard of any of the CEOs. They were all humble people who built teams of excellent colleagues. In fact, the first step in transforming these companies was to get the right people on the bus. They had to develop a staff of very competent team players. The particular skills of the people were not what was critical. What was important was their competence and their ability and willingness to contribute to a team effort.

This may seem to contradict Dr. Deming's injunction to focus on the system not the people. In our opinion there is no contradiction. We are talking about the selection of relatively senior managers. The criterion we are looking for is a willingness to cooperate and learn. These are the people who will build and control the system.

A second thread was finding the right formula. These companies took off when they found a formula that Collins calls a *hedgehog concept*. The hedgehog, a small mammal often confused with the larger and pricklier porcupine, is simple and powerful. When threatened, he rolls up into a compact ball, with spines extended. His defense is difficult to penetrate. Collins's hedgehog concept is depicted in Figure 1.2.

This is the beginning of a system. Each concept interacts with the other two. Combined with a staff of competent team players, this systemic formula enabled these companies to begin achieving phenomenal growth and profitability. According to Collins, it took them an average of three years to find this formula. For example, Walgreens, one of the 11 companies that met Collins's criterion, decided it could be the most *convenient* superdrugstore. It further decided that its economic engine would be *revenue per customer visit*, rather than profit per store. This led Walgreens to do some surprising things. If they had a store in the middle of the block and the corner became available, they would move, because the corner was more convenient. In one area, they built seven stores within a mile of each other. If profit per store

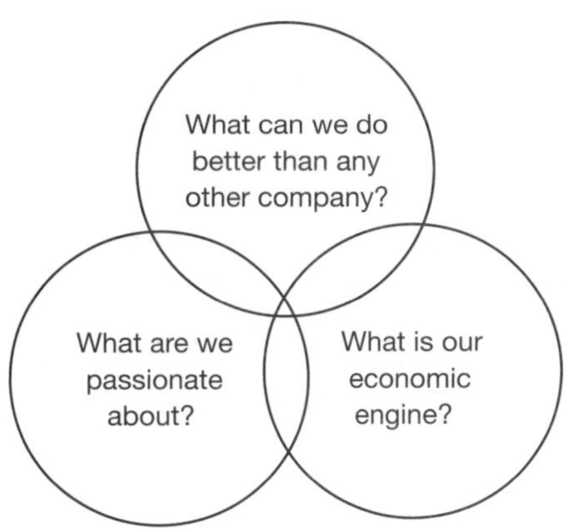

Figure 1.2 The hedgehog concept.

had been their economic engine, they would never have done this. But each of the stores offered convenience to a particular set of customers, and the engine was revenue per customer visit not profit per store.

One final note: Collins used the criterion of 15 years of outstanding performance so that the transformation would have to outlast the initial CEO. Many more companies achieved exceptional performance for shorter periods because of a brilliant CEO but failed to develop a staff and a system that could deliver exceptional performance after the CEO's departure. This emphasizes the point that the system is more important than the people. Although the system must be initially constructed by people, a properly constructed system outlives those who constructed it.

In practice, it is very difficult for most of us to focus on the system rather than the individuals in it. Perhaps this is because we live in a culture that is so focused on the individual. Focus on the system requires great discipline. When some unpleasant event arises in business, we are most apt to ask, "Who is responsible?" It is a more difficult and slower process to figure out what it was in our system that led to the problem. This hits senior managers the hardest. Since they have the most control of the system, they will probably bear the most responsibility for what happened.

An example of looking at the people rather than the system is found in the Enron scandal. The general assumption is that the senior management of the company contained a number of crooks. Upon reflection, we should realize that the Enron fraud was abetted by their auditor, Arthur Andersen, and by their major banks, including J. P. Morgan and Citicorp. All three companies have already paid large fines, based on their complicity. The scandal was certainly abetted by existing rules of financial disclosure. It required the cooperation of many Enron employees as well. Enron and many other financial scandals arose in a system that offers huge incentives to managers for short-term profits, that requires less than adequate financial transparency, and in which a company can fire its auditors and hire new ones if it does not like what the auditors say. We can say that the offending Enron executives should be put in jail, but without systemic change, we are likely to see this problem repeated again and again.

It is important, as a manager, to exercise the discipline to look at the system and avoid jumping to blame individuals. I wrote an article about this need years ago, titled "Kicking the Habit [of Poor Management]" (Carder 1991). In it, I compare the habit of focusing on the individual to an addiction, and I suggest that the methods of 12-step programs could be effective in helping us break the habit. It is a hard habit to break, but the reward is great. If we take the easy shortcut and focus on the individuals, we will be very unlikely to create a lasting improvement.

THE 14 POINTS FOR MANAGEMENT

Over the years, Deming distilled his observations into a set of 14 points for management. These 14 points are based on more than a half century of observing and working with business. According to Dr. Deming, the 14 points have one aim: "to make it possible for people to work with joy." We will briefly discuss each of the points with some reference to its application to safety.[3]

1. *Create constancy of purpose for the improvement of product and service, with the aim to become competitive, stay in business, and provide jobs.* Constancy of purpose is the conceptual framework that gets everyone in the company moving in the same direction. It establishes priorities and guides decision making. With regard to safety, constancy of purpose is established when every person in the company knows that safety is an important priority, and that the ultimate goal of the operation is constant improvement toward zero accidents. We shall see that management cannot establish constancy of purpose for safety just by talking about it. Management must personally demonstrate their commitment to safety in order to establish constancy of purpose.

2. *Adopt the philosophy of cooperation (win-win) in which everybody wins. Put it into practice and teach it to employees, customers, and suppliers.* The notion that business is improved by having one department compete against another is as bankrupt as the notion that a football team would be improved if the linemen competed against the running backs. Safety clearly requires cooperation among employees, departments, management, unions, industry, and regulators.

3. *Cease dependence on mass inspection to achieve quality. Improve the process and build quality into the process in the first place.* This is an exceedingly difficult concept to grasp. The analog in safety to mass inspection is the safety audit. Inspection will not improve quality, because quality comes from the manufacturing process. The inspector may eliminate some poor product but can never eliminate all of it. Even when it is found, it must be discarded or reworked, adding cost. The challenge is to improve the process so that defective products are not made in the first place.

   Similarly, audits and inspections do not create safe processes. At best, they may provide evidence that the process is unsafe. However, correcting the violations is not the answer. The question is what is it about the process that allowed the violations to occur. If the process is not fixed, a new set of violations is sure to develop just as soon as we fix the old ones.

4. *End the practice of awarding business on the basis of price tag alone. Instead, minimize total cost in the long run. Move toward a single supplier for any one item and a long-term relationship of loyalty and trust.* One place where lowest-price bidders can lead to a safety problem is in the area of contractors. A contractor with a poor safety record and unwillingness to cooperate with the customer may be the low bidder but will seldom yield the lowest total cost.

5. *Improve constantly and forever the system of production, planning, or any activity that will improve quality and productivity and thus constantly decrease costs.* No process is ever perfect; it can always be made better. The history of business is rife with the corpses of businesses that thought they had found the best way to do something and stopped looking for better ways. In safety, there is always room for improvement. This is especially true when we view safety as having a direct relationship to our ability to understand and control our processes. A process that is under control and has low variability will be safer and more productive, and will produce a higher quality product.

6. *Institute training for skills.* Most businesses we are familiar with violate a number of Deming's training principles. Most often we see what Dr. Deming called *worker training worker*. On a relatively haphazard basis, the new worker is assigned to a more experienced worker for training. Over successive generations, this method causes ever-increasing variation. The analogy is cutting fence posts by using the last post cut to measure the next one. Just as the fence posts will not be of predictable lengths, so the training will not be predictable content or quality. There should be a reference standard and trainers (who may be workers) who are able to train to that standard. A poorly trained worker is an unsafe worker.

7. *Adopt and institute leadership for the management of people, recognizing their different abilities, capabilities, and aspirations. The aim of leadership should be to help people, machines, and gadgets do a better job. Leadership of management is in need of overhaul, as well as leadership of production workers.* As we shall see later in the book, *management* of safety might be

accomplished by instructing workers to work safely. Leadership requires personally demonstrating commitment to safety. Leaders are judged by what they do as well as by what they say. Leaders should be trained to recognize this as a part of their role and that different people will have different abilities, capabilities, and aspirations about safety. Leaders need to manage with an understanding that not all people are the same and that it is possible to create an effective system that takes advantage of this variation in people, rather than attempting to stifle it.

8. *Drive out fear and build trust so that everyone can work effectively.* Psychological evidence tells us that a number of problems are associated with using fear as a motivator. In particular, fear is especially damaging to cooperation and creativity, which are critical behaviors in the modern workplace. In addition, frightened workers will not tell us about problems or near misses and will often fail to report accidents. Without this information, it is difficult to improve the safety of work processes. Frightened workers will not cooperate effectively with management or with other workers.

9. *Break down barriers between departments. Abolish competition and build a win-win system in the organization.* It is virtually impossible to improve process safety without cooperation among departments like engineering, production, and maintenance. In practice, we very often see competition among them. A profound example of the consequences of failure of departments to cooperate is the Challenger disaster. The investigation of the incident (Feynman 1988), revealed clearly that there was a huge failure of cooperation between engineering and management. The engineers responsible for the flight refused to sign off on it because of the danger posed by the low temperature at the launch site. Management failed to understand or heed their advice.

10. *Eliminate slogans, exhortations, and targets asking for zero defects or new levels of productivity. Such exhortations create only adversarial relationships, as the bulk of causes of low quality and low productivity belong to the system and thus lie beyond the power of the workforce.* A fundamental premise of this book is that the bulk of causes of accidents are also in the system and not in the workforce. This does not mean that the causes of accidents lie only in the physical environment. The system includes the people as well as the procedures and methods

by which they work. It includes planning, training, and education. It includes the policies of the organization, such as the methods by which employees and managers are selected, promoted, and compensated. This assertion is strongly supported by the realization that one plant would not have one-tenth as many accidents as a similar plant just because the former plant has better workers. It has to have a better system.

11. *Eliminate numerical goals, numerical quotas, and management by objectives. Substitute leadership.* The point of this is not that goals are bad. The problem is that goals, by themselves, accomplish nothing. Rather than establishing goals and quotas, the job of leadership is to improve the system. We should point out that some goals are imposed by the external world and are very important. The business that consistently loses money will not survive. Nor is it acceptable for a business to injure or kill people. The only appropriate goal for safety is that no one should be injured. We should not have to spend much time getting agreement on that point; we should move on to the real work of improving the processes in order to prevent accidents.

12. *Remove barriers that rob people of joy in their work. This will mean abolishing the annual rating or merit system that ranks people and creates competition and conflict.* To the extent that intrinsic motivation is present, people will enjoy work. Intrinsic motivation is the motivation that comes from the work itself. Golf is intrinsically motivating for most golfers. It is fun to play. We do not need an incentive to participate. In fact, putting in external incentives can destroy that pleasure. The weekend golfer gets great joy out of a good tee shot. The professional, who must play well to survive, often enjoys it much less. To the extent that workers find joy in their work, they will be energetic and committed. To the extent that they are externally controlled and manipulated, their joy will be reduced, along with their energy and commitment.

    The job of management, then, is not to motivate employees but to work with them to create a system in which everyone can work effectively and contribute to the success of the group. Individuals working in such a system will be motivated. Certainly issues like compensation and evaluation cannot be ignored. However, they are no substitute for building a system that works. In fact, compensation and evaluation are more often roadblocks to motivation than they are sources of it.

In safety, the motivation of the worker is obvious. The notion that workers would frequently injure themselves without the intervention of wise management, which is implied in many safety programs, is insulting and helps diminish the joy in work.

13. *Institute a vigorous program of education and self-improvement.* Education is different from training. Training relates to developing proficiency in a skill. Education is learning theory. Studies of education in industry have repeatedly shown a very high return on investment (Bassi et al. 2003). We have seen numerous cases in which process incidents—chemical reactions that got out of control and led to fires and explosions—could have been prevented by education. Had the process operators had some rudimentary education in chemistry and physics, they would have known that something was very wrong long before disaster occurred. It is not possible to write procedures that can cover every eventuality. Education can equip workers with knowledge that will help in those situations where the procedures do not exist. Without proper education in a subject, how can an individual be expected to perform appropriately?

    Educating management is no less important. In particular, this includes education in areas that Deming called profound knowledge: variation, psychology, and the theory of systems. These areas of knowledge are critical to the effective leadership of process improvement.

14. *Put everybody in the company to work to accomplish the transformation. The transformation is everybody's job.* The fact that management has relatively more control over the system does not mean that it has all of the responsibility or control. Although the system may be the dominant factor in an organization's performance, this does not excuse worker cynicism and refusing to try to do better. Everyone is ultimately responsible for their own behavior. The job of leadership is to *enable* everyone to succeed. Leadership cannot *ensure* anyone's success. A safe workplace can never be achieved without putting everybody to work on the transformation.

APPLYING THE 14 POINTS

THE 14 POINTS CONSTITUTE A SYSTEM. ONE IS NOT FREE TO CHOOSE SOME TO IMPLEMENT AND IGNORE OR CONTRADICT OTHERS.

The 14 points constitute a system. One is not free to choose some to implement and ignore or contradict others. They are highly interdependent. Without *driving out fear*, it is not possible to *put everybody to work to accomplish the transformation. Slogans*, exhortations, numerical goals and quotas are among the *barriers that rob people of joy in work. Dependence on mass inspection* also interferes with *joy in work*. Without *constancy of purpose,* very little can happen at all. If you find a point that is contrary to your beliefs, values, or policies, you will need to examine and understand the impact of your position on the remaining 13 points. You should not try to install a system that is missing a critical part. You may be better off not trying to implement the other 13.

The biggest conflict usually involves the annual merit review. Deming was not against reviewing employee's work. He did have serious questions about the ability of most managers to evaluate the performance of employees accurately, because that performance is the output of the employee working in the system rather than of the employee alone. An individual working in an inefficient system and getting moderate results might actually be doing a better job than another working in a highly effective system getting better results. Deming's alternative to rating employees formally was recommending that managers have an extended conversation/dialogue with each subordinate annually or more frequently. Such a discussion would cover the perceived performance of the employee and the potential systemic barriers to improving that performance.

However, the biggest problem is created when the annual review pits one employee against another in a form of competition. This creates fear and reduces cooperation, which impair safety and productivity. Alfie Kohn, in his book *No Contest* (Kohn 1986), has written very eloquently on the dangers of introducing competition into business. Moreover, he cites a number of studies demonstrating that the most successful scientists and the most successful business leaders are not as competitive as their less successful colleagues. Competition among individuals does not always lead to better performance or better products.

We sometimes use the 14 points as a diagnostic tool. After explaining the points to a group of employees, we ask them to fill out three sticky notes, listing the three points they feel the company most needs to work on at the present time, one on each note. No matter how excellent your company may be, the system needs to be improved constantly and forever. There is always room for improvement.

The responses of the group can be posted on the wall in the form of a bar chart or Pareto chart, as in Figure 1.3.

The next step is to begin to discuss the reasons people made the choices they did. The exercise serves two purposes. First, it helps teach people the meaning of the 14 points. In order to select the points that are most relevant to the organization, the employees have to think about them. Second, it provides insight into the dynamics of the organization and may help identify a starting point for improvement efforts. Although the 14 points constitute an interrelated system, you cannot work on all of them at the same time. You must focus your improvements on areas that will have the greatest impact or leverage on the rest of the system. This exercise can provide us with insights about leverage.

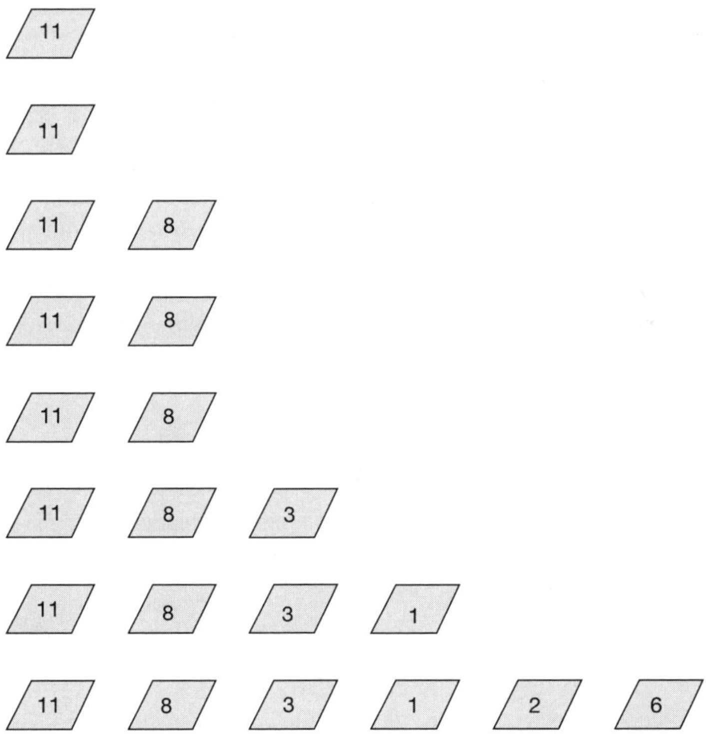

Figure 1.3 The employees' selection of points most important to their company.

One of the important reasons to conduct this type of exercise is that it provides a very comprehensive picture. It provides a view of the entire system from a very high altitude (meaning that we see the whole system but with very little detail, as in the view of a landscape from an airplane.) You may not see much detail, but everything is included. We spend so much of our time working on details that we often miss the forest for the trees. This is especially true for the safety professional who must deal with so many issues of regulation and compliance that it is easy to lose sight of the organizational system that controls the safety of work.

The 14 points are a set of principles for operating a highly safe and productive business. It is worth noting that not one of the 14 points suggests that safety should ever be compromised. If it did, this would diminish joy in work, increase fear and distrust, limit quality and productivity, and certainly run contrary to the new philosophy. Safety is inherent in the 14 points in the same way that quality and productivity are inherent.

KEY POINTS

- Ultimately this book is about is how to create positive change in organizations and individuals. Our challenge is to find leverage for change.
- Dr. Deming's approach to management is founded on two fundamental principles. Although based in science, these principles form the philosophical framework of his system:

  Principle 1. People will do their best to do what they think is expected of them.

  Principle 2. To improve performance you need to improve the system rather than focus on the individuals.

- Dr. Deming offered a system of 14 points for management, based on over 50 years of experience helping businesses improve quality, productivity, and economic success:

  1. Create constancy of purpose for the improvement of product and service, with the aim to become competitive, stay in business, and provide jobs.
  2. Adopt the philosophy of cooperation (win-win) in which everybody wins. Put it into practice and teach it to employees, customers, and suppliers.

3. Cease dependence on mass inspection to achieve quality.
4. End the practice of awarding business on the basis of price tag alone.
5. Improve constantly and forever the system of production, planning, or any activity that will improve quality and productivity and thus constantly decrease costs.
6. Institute training for skills.
7. Adopt and institute leadership for the management of people, recognizing their different abilities, capabilities, and aspirations. The aim of leadership should be to help people, machines, and gadgets do a better job. Leadership of management is in need of overhaul, as well as leadership of production workers.
8. Drive out fear and build trust so that everyone can work effectively.
9. Break down barriers between departments. Abolish competition and build a win-win system in the organization.
10. Eliminate slogans, exhortations, and targets asking for zero defects or new levels of productivity. Such exhortations create only adversarial relationships, as the bulk of causes of low quality and low productivity belong to the system and thus lie beyond the power of the workforce.
11. Eliminate numerical goals, numerical quotas, and management by objectives. Substitute leadership.
12. Remove barriers that rob people of joy in their work. This will mean abolishing the annual rating or merit system that ranks people and creates competition and conflict.
13. Institute a vigorous program of education and self-improvement.
14. Put everybody in the company to work to accomplish the transformation. The transformation is everybody's job.

- The 14 points constitute an interrelated system. You cannot pick and choose which points you would like to implement and which you would like to ignore. All are needed together.

ENDNOTES

1. Lean manufacturing is the name given to the process used by Toyota that has provided the highest quality automobiles at very competitive costs.
2. Six Sigma is the name of a quality process developed by Motorola and embraced by many companies, including GE. The name implies a very high standard for eliminating defects.
3. This version of the 14 points was based on discussions between Lloyd Dobbins and Dr. Deming, and was published in *Quality Progress* shortly after Dr. Deming's death.

2

Dr. Deming's System of Profound Knowledge

Dr. Deming's system of management requires applying what he called *profound knowledge* to business problems. In his latter years, Deming talked increasingly about this in his seminars. Profound knowledge is scientific knowledge: theory that has been tested through rigorous observation.

PROFOUND KNOWLEDGE COMPRISES FOUR DISCIPLINES THAT DEMING FELT WERE FUNDAMENTALLY IMPORTANT TO BUSINESS:

1. STATISTICS, OR THE STUDY OF VARIATION
2. PSYCHOLOGY
3. THE THEORY OF KNOWLEDGE
4. THE THEORY OF SYSTEMS

Profound knowledge comprises four disciplines that Deming felt were fundamentally important to business: statistics, or the mathematics of variation; psychology; the theory of knowledge; and the theory of systems. We shall treat each of these briefly in order to illustrate the relevance of each to management.

VARIATION

VARIATION IS A PROPERTY OF ALL MEASURED PHENOMENA. WITHOUT AN UNDERSTANDING OF THE LAWS OF VARIATION, IT IS LIKELY THAT MANY INAPPROPRIATE CONCLUSIONS WILL BE DRAWN FROM THE DATA.

The mathematics of variation is called *statistics*. Some knowledge of variation is essential to conducting any process of measurement, because variation is a property of all measured phenomena. Without an understanding of the laws of variation, it is likely that many inappropriate conclusions will be drawn from the data.

As an example of the importance of knowledge of variation, let us consider two similar chemical plants, each with 50 employees working 2,000 hours each during the year. In 1994, the Riverton plant had two recordable accidents, for a recordable accident rate[1] of 4. The Hillton plant had four recordables, for a rate of 8. Obviously, the Riverton plant has a better safety process, does it not? The answer is that it may, but the evidence just provided is very weak.

Deming explained that accidents should vary according to a *Poisson distribution* (Deming 1993). This is what we expect if accidents are independent of each other and if the probability of an accident's occurring in one period of time is the same as in any other period. Although we could argue with either of these assumptions, they are a reasonable and useful approximation. (One place this breaks down is if an event causes multiple injuries. Poisson distribution applies to events not to total number of injuries.)

The appropriate statistic to calculate the standard deviation of incident rates is U, based on the Poisson distribution. If we use this statistic to compute the expected variation, we find that for the Riverton plant, with an average of two incidents per year, the standard deviation is expressed as:

$$U = \sqrt{\frac{R}{N}}$$

where R is the recordable accident rate and N is the number of units of exposure. In this case, a unit of 200,000 man hours is the approximate annual hours worked by 100 employees. Calculating this for Riverton, with a rate of 4 and 100,000 hours worked for the year, gives us:

$$U = \sqrt{\frac{4}{(100{,}000 / 200{,}000)}} = \sqrt{\frac{4}{.5}} = \sqrt{8} = 2.82$$

The standard deviation is 2.82. Hillton's rate of 8 is only a little more than one standard deviation higher than the rate at Riverton. This is depicted in Figure 2.1.

The chart shows that the Hillton rate is about 1 standard deviation unit above the Riverton rate. A difference this large has a high likelihood of arising by chance. Therefore, one should refrain from taking an action based on the assumption that they are different. One way to refine our measurement

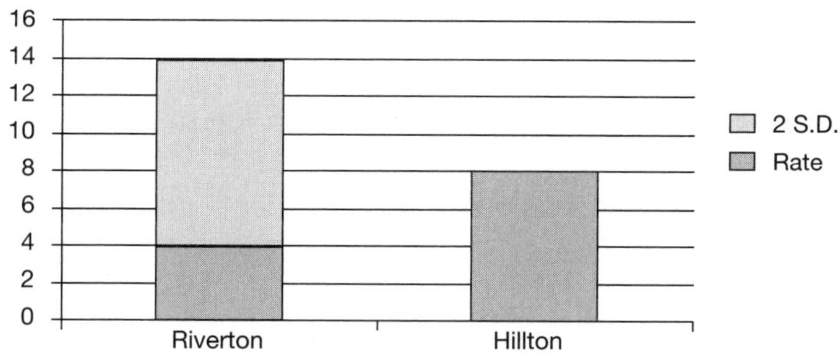

Figure 2.1 Two plants compared, based on one year of data.

is to look at a longer period. When we look at the past 10 years, we find that Riverton has an average rate of 4, and the Hillton plant has an average rate of 8. Computing that standard deviation, *U*, for Riverton using a 10-year period, we have:

$$U = \sqrt{\frac{4}{5}} = \sqrt{.8} = .89$$

Figure 2.2 depicts this.

Because we have more units of exposure, the standard deviation becomes smaller. Now we see that Hillton has a rate that is over 4 standard deviations higher than Riverton over the 10-year period. Although this could still be a chance result, it is very unlikely, with a probability of less than one in 1,000. It would be appropriate, based on these data, to conclude that Riverton is indeed superior, *if you were confident that each plant is recording its accidents with equivalent methods and standards.*

We have to be very cautious about judging management performance on the basis of accident rates. Doing so can create a strong incentive to modify criteria in order to report fewer accidents. We have seen this range from outright fraud to subtle or not so subtle messages that cause workers to conceal minor incidents. In one large plant (1,200 employees) we visited, the recordable rate had gone from a long-term rate of 11 down to 7 in a single quarter. This is a change of more than 3 standard deviation units and, therefore, is very unlikely to have been a result of chance variation. When we asked how this was achieved, the plant manager told us he had instituted a rule requiring that all reports of recordable accidents be made in person to him. He told us this enabled him, presumably because of his extreme competence, to identify root causes and eliminate them, thereby improving the system. Although the plant manager was confident that his new policy had not caused employees

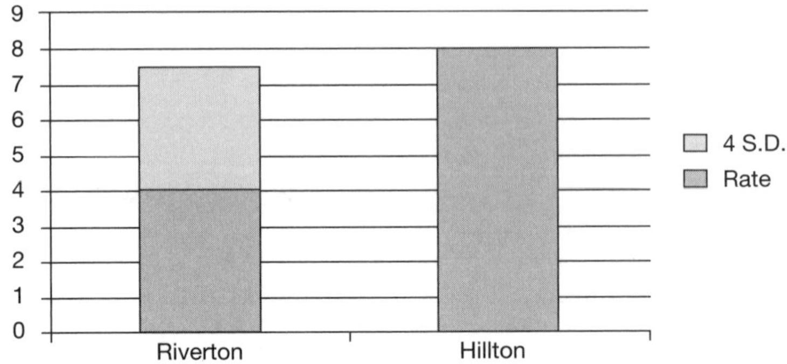

Figure 2.2 The same two plants with 10 years of data.

to conceal accidents to avoid facing him, the employees were at least as certain that some accidents were now going unreported.

When a plant has an unusually high number of accidents in a period, it is common for management to ask what is wrong or suggest that some measures be taken to improve the situation. Without statistical analysis, it is not possible for management to know if anything has changed. Most of the time the unusual periods represent nothing more than random variation. Therefore, most of these communications are counterproductive. At best, they waste time, and at worst, they cause ill-advised actions on the part of local management. Deming called this kind of management intervention that ignores the laws of variation *tampering*. In manufacturing settings, tampering increases the variability of the system and reduces the quality of the product.

PSYCHOLOGY

Dr. Deming was trained as a physicist. Many people believe that his approach to quality improvement was through the window of statistics, and initially it was. However, in later years, he became increasingly interested in psychology. In a four-day seminar in San Diego in 1988, Dr. Deming learned that several attendees held PhDs in psychology. He arranged a session in which he was the student and the assembled PhDs applied their special knowledge to his work. This process subsequently became institutionalized as a part of his four-day seminars. An evening would be set aside for an optional session. Deming would sit in the audience, and the session would be facilitated by a psychologist. The focus would be on what Deming called the *psychology of change*.

BECAUSE MANAGERS GENERALLY MANAGE PEOPLE, THEY ARE NECESSARILY PRACTITIONERS OF PSYCHOLOGY.

Because managers generally manage people, they are necessarily practitioners of psychology. Typically, they have little or no training in this discipline. Moreover, much of what constitutes common knowledge about psychology is wrong. Use of this common knowledge often causes serious harm.

Perhaps the greatest misconception about psychology is how incentives function. From the 1940s through the 1970s, behaviorism was a prominent school of psychology, exemplified by Skinner and his students. Behaviorism was anchored by the belief that incentives, which they termed *positive reinforcement*, were the fundamental mechanism for controlling behavior. Behaviors that led to receiving incentives would be strengthened and more likely to occur in the future. Skinner showed that even complex behaviors could be developed through the use of positive reinforcement by using a process called *shaping*. In shaping, the subject receives reinforcement at first for a small component of the desired behavior. When the subject learns to do that, then an additional component is added. The subject learns successively closer approximations to the desired behavior until finally it has developed a very complex behavior. Using this method, Skinner taught pigeons to play table tennis and to guide missiles. (The latter was rejected by generals during World War II, in spite of promising demonstrations of its feasibility.)

Although incentives clearly influence behavior, it is equally evident from the literature of psychology that incentives are in no way the sole mechanism of behavioral control. In fact, they often do the opposite of what we would expect. For an excellent discussion of some of the problems of incentives, see Alfie Kohn's *Punished by Rewards* (1995). Kohn notes, for example, that many studies have demonstrated that positive reinforcement actually undermines intrinsic motivation. If a person enjoys doing something, then paying the person to do it will actually make the activity less attractive. Many professional athletes describe a similar phenomenon of seeing the fun go out of their sport when they play for money. Intrinsic motivation, or the joy of doing something for its own sake, is a strong motivator. Kohn cites over 70 published studies demonstrating that incentives interfere with intrinsic motivation.

Incentives often do produce the desired behavior, but at the same time may fail to change attitudes and beliefs that underlie the behavior. In addition, the incentive may change behavior in a manner that is much different than management intends. A simple example is found in programs in which a strong incentive is provided for not having any accidents. A common result of this is nonreporting. Of course, management may respond by establishing an incentive for accurate reporting. This kind of response increases the complexity of reward systems which may make them so confusing as to be totally ineffective.

Punishment creates more problems than positive incentives do. Even the behaviorists argued that punishment is generally ineffective. One problem is that punishment creates fear, and fear is not compatible with learning, creativity, or cooperation. Punishment will certainly establish superficial compliance. If individuals are strongly punished for breaking rules, they will definitely attempt to avoid being caught breaking these rules. Surprisingly enough, the punishment may reduce their conviction that the rule is important.

Incentives and Attitudes

Although incentives often exert influence over behavior, they do not necessarily impact attitudes the way one would expect. It is easy to pay people to do particular jobs, but we cannot use pay to get them to like the job or the company they work for. In fact, increasing the pay often decreases the employee's loyalty to the company. To understand this, we turn to the theory of *cognitive dissonance*. Cognitive dissonance (Festinger 1957) is a theory of attitude change that was developed to explain some experiments that initially appeared to give counterintuitive results. In a classic study, students were asked to give a speech supporting a position contrary to their own political view. Some subjects were given a relatively large amount of money to do this, while others were given very little. The experimenters then measured changes in the student's underlying attitude toward the position they were asked to advocate. Surprisingly, the students who were paid the least showed the most change.

WHEN PEOPLE HAVE TWO THOUGHTS (COGNITIONS) THAT ARE INCOMPATIBLE (DISSONANT), THEY WILL CHANGE THEIR ATTITUDE IN ORDER TO MAKE THE THOUGHTS LESS DISSONANT.

The experimenters reasoned that students who advocated a position they did not believe in, but who were paid well, could rationalize the speech by saying they did it because of the money. Those who were paid very little did not have this rationalization. Instead, they had to rationalize by deciding that the position was not really so bad, thereby changing their attitude in the direction of that position. The theory states that when people have two thoughts (cognitions) that are incompatible (dissonant), they will change their attitude in order to make the thoughts less dissonant. In the case of the political speech, they have two incompatible cognitions: "I am a Democrat," and "I gave a Republican speech." If they are paid enough, they can say, "I just did it for the money." If not, they are more likely to say, "Well, the Republican position is not really that bad."

Organizations that use strong incentives and punishment to enforce rules and policies may get extensive compliance but little commitment. The employees may feel they are following the rules to avoid severe consequences not because the rules are good rules and intrinsically worthy of being followed. This leads to following the letter of the rule not the spirit of the rule. It can also lead to breaking the rules if employees can be certain they will not be caught.

In our work with the survey process discussed in Chapter 8, we learned about the employee's perception of rules and enforcement through the results of questions about off-the-job safety. Sites that were shown by these questions to have effective off-the-job safety programs tended to have excellent safety performance. In our focus groups, employees told us that these programs distinguished between rule-based safety and belief-based safety. If the plant had an off-the-job safety program, it indicated that management believed safety was the right thing to do rather than the expedient thing to do.

We do not argue that discipline is not appropriate or that incentives and punishment should never be used. There are certainly circumstances in which punishment is necessary. It is definitely appropriate when individuals are gaining personal advantage at the expense of other employees by violating rules. In our experience, however, this is not usually the case. More often employees violate rules with the intention of getting the job done. Frequently they do so with the implicit or explicit consent of supervision and management. In these cases, punishment of the employee is seldom appropriate. Rather, it is important to understand why the employees feel it is better to violate the rules. Do the rules make sense? What message is being sent by management? Do the employees believe they should take risks in order to get the job done? Why? Are supervisors and managers supporting this risk-taking behavior by praising them for work done on time without stopping unsafe work practices that merely get the job done?

Proper use of incentives presents a very complex problem. It makes sense that people who do a better job should get paid more. A problem with this is that it is often difficult or impossible to reliably determine exactly who was doing a better job. Deming's solution was to base pay on such things as education and time of service. In our own opinion, this is not sufficient. However, incentives should be used only when one has carefully looked at the problems and conflicts that the incentives might cause. For example, there is no evidence that the huge incentives provided CEOs for company performance have been effective. The 11 CEOs in Collins's study who led profound transformations were not paid nearly as much as many of their less successful colleagues. On the other hand, some of the highest paid CEOs in the last decade presided over massive frauds and led companies like Enron and WorldCom into bankruptcy. This happened in a system in which these executives received huge profits from stock options and compensation plans based on increasing the stock price. There is only a difference of scale between workers failing to report an accident in order to get bonuses and executives cooking the books to get their bonuses. Both are driven, in part, by a defective incentive system.

Another complexity of incentives is that a particular incentive may have a very different value to different persons. For example, one company we worked with offered an opportunity to make a presentation to the executive committee as a safety incentive. This was attractive to most managers, but when the same incentive was offered to hourly employees, they found it very unattractive. The hourly employees and some of the managers were very uncomfortable with making presentations, especially to high-level groups, such as the executive committee. Several very good safety programs in the company went unrecognized because their leaders would not submit the site names for recognition for fear of having to give this executive committee presentation. Unfortunately this opportunity was such a strong incentive for others that some less than stellar safety programs were built up to more than they were, just to get a chance to speak in front of the executive committee.

Perception and Memory

> PERCEPTION IS *CONSTRUCTED* FROM INFORMATION RECEIVED (FROM ALL OF THE SENSES). MEMORY IS THE RECORD OF THAT PERCEPTION OF REALITY, NOT THE REALITY ITSELF.

Contrary to widespread belief, perception and memory do not work like a video camera and VCR. We do not create an endless video and audio record in our head that can be accessed completely under hypnosis or some other special condition. Studies of both perception and memory make it clear that both processes involve construction not just reception. The optic system transmits a great deal of information about the environment to the brain, but the perception is *constructed* from this information. The perception may be quite accurate, or it may be quite inaccurate when referenced against some standard. The perception is based not only on what information is coming in through the optic system but also on past experience. We tend to see and hear what we expect to see and hear. This is not because we consciously bend the truth but rather because the constructive process is influenced by our past experience, our future expectations, and our beliefs about the world and how it works. For example, new scientific theories are rejected as absurd until they are widely adopted. Then no one can understand how we believed anything else.

In 1982, the Australian scientists Barry Marshall and Robin Warren (Marshall 2002) discovered the bacterium *helicobacter pylori* in the stomachs of ulcer patients. For over 10 years, with strong scientific evidence behind them, they attempted to convince physicians that ulcers were a result of bacterial infection. They were told this was absurd. Bacteria could not live in the strong acidity of the stomach. In the course of their work, they found that previous studies had also discovered bacteria in the stomachs of ulcer patients. However, because it was generally accepted that the environment was too acidic for bacteria to survive, these previous studies assumed that the bacteria had been the result of some contamination of their samples. They constructed a perception of reality based on their beliefs.

Nearly 20 years after being first published, Marshall and Warren's theory achieved general acceptance. They showed that *helicobacter pylori* are able to burrow into the mucosa lining of the stomach and were thereby protected from the acidity. This is of no small consequence, since it indicates that treating ulcers with a regimen that includes antibiotics will have a better therapeutic effect than the prevalent treatment with acid blockers alone. In fact, using antibiotics vastly increases the success rate of ulcer treatment. The acid blockers may actually create an environment more hospitable to the bacteria than no treatment at all. The acid blockers may be treating the symptom but aiding the root cause of the condition. The point here is that previous studies had observed the presence of bacteria in ulcer patients but had rejected or ignored it. They did not expect bacteria there, so they decided that it was not there, rejecting their own data.

In a similar vein, many studies have shown that Democrats grasp the salient points in their candidate's arguments but more often fail to hear or understand points made by Republican speakers. Similarly, Republicans grasp the points made by their party members but fail to hear and understand those made by Democrats.

The applications of this theory to safety management are quite considerable. We are all familiar with individuals willing to takes risks because they are convinced they will not be injured. Such people are underestimating the probability of an unfavorable event by paying attention to past success with risk and ignoring the possibility of an accident and its consequences. Of course, the same process applies to managers who take rather more substantial risks on the basis of their own underestimate of the probability of unfavorable outcomes.

Hearing and seeing what we expect or what we want to occur can have a powerful effect on safety. We can see this at work in the space-shuttle accidents. In an outstanding example of incident investigation, Richard Feynman (1988), a Nobel laureate and the leading physicist of his day, recounted his experience while serving on the presidential commission that investigated the Challenger explosion. At one point, Feynman polled a number of engineers, asking them about the probability of catastrophic failure on a shuttle mission. The consistent estimate was on the order of 1 in 100. When the same question was posed to managers, individuals who would ultimately be held responsible for a catastrophic failure, the consistent answer was in the range of 1 in 1,000. Feynman's own estimates, based on extensive discussions with engineers and review of reliability data from prior flights, was on the order of 1 in 100. The managers were understating the probability of an unfavorable outcome, in this case with tragic consequences. In fact, up to and including the Columbia disaster, there have been two catastrophic failures in 113 launches. This is strong evidence that Feynman's estimate (and that of the engineers) was of the appropriate order of magnitude.

Our discussion of cognitive dissonance might give us a potential remedy for this. If we could induce workers and managers to discuss what can go wrong, and in fact advocate the position that an unfavorable outcome would occur, they would be more likely to entertain the idea that an unfavorable outcome (accident) would occur and might modify their behavior to prevent it. In fact this may be the reason for the success of the job safety analysis (JSA), in which workers discuss potential hazards of a task before it is undertaken.

The U.S. Army has undertaken a process similar to JSA, which they call *risk management*. In this case, it is officers who analyze potential hazards. The Army's data clearly demonstrate a reduction in injuries during infantry training as a result of this risk-management process. A similar process is widely used in quality management, called failure mode and effects analysis (FMEA). This involves a detailed engineering analysis of how a part or a process might fail, including an estimation of the probability of failure and the likely consequences of such failure. Through this focus, the designers are likely to get a much more realistic picture of the potential problems of a system.

Of course, the implications of this for safety managers are clear. To tell line managers about hazards may have minimal impact. Although management may agree that the hazard is real, they may fail to hear the safety mangers' arguments and may substantially underestimate the probability of an accident. The solution is not to make a stronger case but rather to induce the line managers to analyze hazards themselves. By having them make the case, we should see a rise in their estimate of the probability and ultimate risk[2] of an accident.

THEORY OF KNOWLEDGE

The theory of knowledge, related to the philosophy of science, is critical to Dr. Deming's approach to business. It defines the rules of how knowledge is acquired. Deming asserts, quite properly, that knowledge is based on theory. A theory is a model of reality that allows us to predict outcomes. To qualify as a theory, the model must make testable predictions, and the predictions must be capable of being wrong. That is why after-the-fact explanations are not necessarily theories. They are only theories if they make future predictions that can be tested.

AFTER-THE-FACT EXPLANATIONS ARE NOT NECESSARILY THEORIES. THEY ARE ONLY THEORIES IF THEY MAKE FUTURE PREDICTIONS THAT CAN BE TESTED.

Psychoanalytic theory, for example, is not a theory. Any conceivable thought or behavior can be explained by it. There is no way that the validity of the theory can be put to a rigorous test. It may have entertainment value, and it may possibly offer comfort and healing to patients of psychotherapy, but it does not qualify as a scientific theory.

On the other hand, some of what Heinrich said about safety does indeed constitute a testable theory. Heinrich depicted the relationship between minor incidents and major accidents in the form of a triangle, similar to that depicted in Figure 2.3. He asserted that severity of accidents was largely fortuitous and that, therefore, frequency and severity should bear a constant relationship. That is clearly not the case (Manuele 2003; Petersen 2000). Although it means that Heinrich was wrong about this particular, at least he presented us with a testable theory. Unfortunately, the evidence that he is wrong has not dissuaded most organizations from carefully investigating every minor recordable incident under the mistaken assumption that these incidents bear a direct relationship to catastrophic problems. At the same time, these organizations ignore near hits/misses that portend terrible accidents. We do not assert that the particular ratios shown in the diagram are accurate. But it does graphically represent the concept that too much time is spent on minor incidents that will never result in a major or serious accident while much less time is spent on some near hits/misses that should be investigated in detail. Does a paper cut really warrant more of an investigation than the proverbial beam that fell near, but not on, the employee?

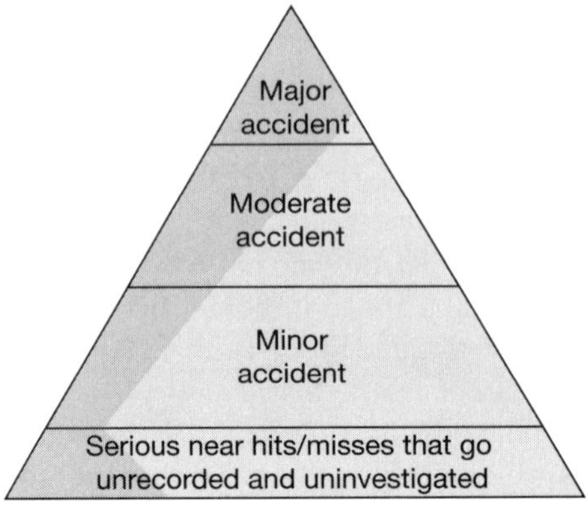

Figure 2.3 Revised theoretical relationship of minor and moderate accidents that have the potential to yield major accidents.

Does a burn from brushing against an uninsulated steam tracer really warrant more of an investigation than a process-vessel overpressurization that results in relieving hazardous material to a contained vent system? Are we spending our limited resources where they will have the most benefit? Or are we spending these resources where we can because we do not understand the accident-producing processes well enough to better focus our efforts?

One of the conclusions to be drawn from the assertion that knowledge requires theory is that it would make sense to teach theory to workers. A worker armed with an extensive set of rules and procedures does not necessarily have the tools necessary for safe work. These procedures and rules do not constitute knowledge. A worker operating a chemical manufacturing process would be much more capable of safe operation with some knowledge (theory) of the reaction taking place, the process controlling it, and the measuring devices monitoring it. Rules and procedures can never cover every eventuality. Moreover, the more extensive and complex the rules and procedures, the less likely they will be known, understood, and followed. Learning of theory would allow the worker to make predictions if something unforeseen happened. We are aware of numerous instances in which workers followed the rules and procedures but, nevertheless, suffered serious injury. From our review of the events leading up to the injuries, we are convinced that had the workers had a better understanding of the process they were operating, they would have been able to avoid injury. We need to better use the strengths of human and nonhuman controls. Nonhuman controls, computers, relief valves, level gauges, and so on, are very good at vigilantly following the rules they are set up to follow (programmed instruction, relief-pressure setting, level setting) and initiating action as a result of those rules. Humans are not nearly as good as computers and machines at following rules and procedures. On the other hand, nonhuman controls will only work if we know all of the possible conditions and how to deal with them. Humans, when properly trained and educated, are very good at reacting appropriately to conditions that were not anticipated by the programmers of nonhuman controls. Humans make great humans. Perhaps it would be better to let machines do more of watching routine processes and spend that human time training and learning how to react to the alarms or nonroutine events when they inevitably occur.

As an example, an operator in a chemical plant was required to place reagents in a vessel and heat them to a specified temperature according to a gage on the vessel. The gage stopped at a temperature below the specification and failed to rise in spite of the continuous application of heat. To someone with a basic knowledge of physics, this suggests a serious problem. Either the heat source is defective, the gage is defective, or something is dissipating heat, which could include an undesired reaction or a leak in the vessel. Any of these would be a cause for immediately stopping. The operator, lacking this knowledge, pondered the situation for a while. Finally he fled just before the vessel exploded, indicating that he did have at least some knowledge. The temperature gage had failed.

If you ask the question, "Do the workers know how to do the job?" you should consider the meaning of the word "know" in this context. They may remember the procedures, but do they have any theory? If you avoid teaching the theory, you need to assume that you can write procedures to cover every possible event.

A critical component of the theory of knowledge is the *operational definition*. An operational definition defines a construct by providing sufficient information to measure it. For example, hunger is a feeling most of us have experienced. However, if we want to study hunger scientifically, we need to define it operationally. We might define it through self-reports, such as "my hunger is 7 on a 10-point scale." We might define it as the amount of time since the last meal. We might define it as the quantity of ice cream that a person would eat in 30 minutes. In each case, we would need to be careful. What kind of ice cream? Was the last meal of sufficient quantity to satiate the person?

If anyone using our definition can make the same measurement and get a very similar result, we have met the criterion of operational definition. However, operational definition is only a first step. Just because we have an operational definition that will enable measurement does not mean we have a measure that will give us insights that lead to improving the system.

THEORY OF SYSTEMS

Deming called his explication of business as a system "the spark that ignited Japan." The crucial insight Japanese executives got from Deming was that a business is a system of interrelated parts, and that the crucial function of management is to manage the whole rather than attempting to manage the parts separately.

It follows from this, and from Deming's assertion that most problems lie in the system rather than in individuals, that we should consider the theory of systems. This is a formal discipline that has received considerable attention since the publication of Peter Senge's *The Fifth Discipline* (1990).

Definition of a System

A SYSTEM IS A SET OF INTERRELATED COMPONENTS THAT WORK TOGETHER TO PERFORM A FUNCTION.

A *system* is a set of interrelated components that work together to perform a function. No part in the system can perform the function of the system on its own. The output of the system is not just the arithmatic sum of the outputs of the parts but is a result of the interaction of the parts as well. If interaction does not matter, you do not have a system. An excellent example of a system is an automobile. It is an assembly of several thousand parts that work together to provide a transportation function. None of the individual parts will move you. Many of these parts are required to work together as a system in order to move you. If you take out the fuel pump, the drive shaft, or the ignition system, the car will not work. If you remove nonessential parts, such as the radio, the windows, or the airbags, the car will still provide transportation, but the quality of that transportation will be altered.

Collections of parts do not necessarily make a system. A football team is highly integrated. The runners cannot run if the blockers do not block. The receivers cannot get the ball if the quarterback does not throw to the right place. However, a golf team is integrated little or not at all. The score posted by a golf team (in a medal play event) is equal to the arithmatic sum of the players' scores. Although perhaps the players are able to coach and encourage each other, their interdependence is far less important than that of a football team.

The whole is not equal to the sum of the parts. The quality of an automobile or any other complex system is determined not just by the quality of the parts but by how well they work together. Dr. Deming liked to tell a story about a very good furniture company that decided they should also make pianos. They bought a Steinway piano, disassembled it, and copied each part with great precision. When they assembled their parts, the piano would not play. After more than a year of fruitless effort, they gave up. They reassembled the Steinway piano. *That would not play either.* The moral of the story is that they did not understand the piano as a system. Even with a set of parts that was apparently perfect, they did not have enough understanding of how to make the parts work together properly. This example embodies the first principle of systems theory: the whole is not equal to the sum of the parts.

Optimizing the Parts Does Not Optimize the Whole

Although it does seem logical that the team with the best players, the car with the best parts, or the college with the best faculty would be the best, systems theory tells us this is not true. At the simplest level, consider that the function of the system is not necessarily the same as the function of the parts in the system. Certainly no part of a car will transport you. But imagine a car with the best engine, the best seats, the best sound system, the best climate control system, the best transmission, the best safety features, the best electrical system, the best styling, the best suspension, and so on. Of course, best might be very hard to define. Assuming that we could, however, we might imagine ending up with a car about the size of a Boeing 747.

This is not idle speculation. At a conference some years ago, we heard an automobile executive describe the allocation of electrical power in an automobile that was being designed. The group designing the lighting wanted a large amount of power, as did the group designing climate control, the group designing the sound system, and so on. There was no way to meet the requests of each of the design groups with any reasonable electrical system. Each group wanted to optimize the part of the system they were responsible for. Of course, the job of the executive was to allocate the available power to the various groups in such a way that the resulting car would be optimized.

Good automobiles are a set of compromises. The most important thing is that the parts work together as a system. The quality of an automobile relates to how it drives, how it feels, its reliability, and its value for the money. The Honda Accord, the Lexus ES430, the Toyota Camry, and the Porsche 911 are all fine automobiles. None is distinguished as much by the quality of the individual parts as by how the car works and feels. The parts must be good, of course, but the most important thing is that they be properly integrated into the whole.

Great teams, like great automobiles, are a set of compromises. Basketball is a team sport. The interaction of the players is very important. For every scorer, the team needs a passer. Great offense cannot succeed without good defense. Rebounding requires one person to get the ball and everyone else to screen out the opposition. Optimizing the performance of an individual player or players is not the way to optimize the performance of the team. If we look at the nine highest one-season scoring averages in the history of the National Basketball Association, not one was achieved by a player whose team won the championship in the same year! The Boston Celtics, who won eight consecutive championships and 11 in 13 years, never had a league-leading scorer during that period. Although Michael Jordan was the league's leading scorer in the Bulls' championship years, none of his averages are in the top nine of all time. Moreover, his ability to lead and bring out the best in his teammates was as essential to those championships as his scoring. When Wilt Chamberlain was a prodigious scorer, his teams did not win championships. When he became a better team player and scored much less, his teams were able to win.

Let us consider the safety system. Without trying to define it completely, let us say that it includes many processes, such as safety regulation, enforcement of the regulations, incident investigation, safety training, involvement of employees in safety improvement, and incentive systems for safe work. It involves virtually every structural unit in the company, from executive management and their staff to the most recently recruited hourly workers. It involves human resources, engineering, quality control, and any other group you can name. It includes external groups, most obviously contractors, suppliers, and government. Even the stockholders have an impact, from demands for increased productivity to concerns about the company's public image.

Figure 2.4 represents a way of diagramming the safety system. This diagram has two components: architecture and culture. The architecture represents things that can be changed relatively quickly (although perhaps at great expense). It ranges from the wording of procedures to the design of plants, from who works for the company to the reporting structure of the organization. Any of these things can be changed fairly quickly.

The cultural system represents the attitudes, beliefs, and expectations of the people in the organization. These are the products of experience over time and are relatively resistant to change. All actions of management are interpreted through the lens of this culture. Although the architecture can be changed, the system may not change very much. For example, in organizations that have a history of short-lived initiatives that are soon replaced by new initiatives, employees come to expect the flavor of the month. Thus, any new initiative is greeted with cynicism and resistance that will be passive at best and active at worst. Change is stalled and the expectation of short-lived initiatives becomes a self-fulfilling prophecy. They seem to fail and are replaced again and again.

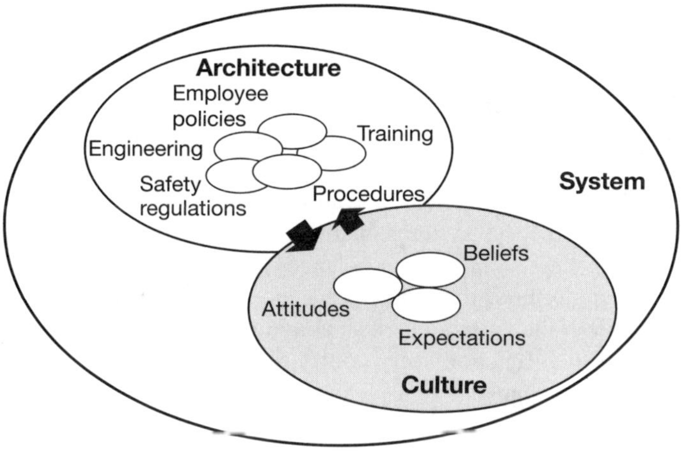

Figure 2.4 The safety system.

ATTEMPTS TO OPTIMIZE THE SYSTEM BY ISOLATING AND OPTIMIZING EACH PART ARE NOT LIKELY TO SUCCEED.

Attempts to optimize the system by isolating and optimizing each part are not likely to succeed. For example, let us say that we want to optimize each process separately. We might begin with strong enforcement of regulations. Next we can work to improve accident investigation. Unfortunately, our enforcement of regulations may interfere with incident investigation and with process improvement. Employees may be afraid to reveal what really goes on, lest they be found to have violated some regulation and be punished for it.

Now let us establish very strong incentives for safe work. Again, this can interfere with incident investigation and process improvement, as individuals become reluctant to report incidents, because this loses them the incentive. Accidents that are not reported cannot be investigated. Therefore, the conditions that produced these accidents may go undiscovered until a very serious event occurs.

None of this is an argument against enforcing regulations or establishing incentives for safe work. What is clearly needed is a program that balances all of these factors. When developing any initiative, we must consider the impact of that initiative on the entire system. Optimizing any particular aspect of the system runs the risk of driving the system out of balance. Things will not improve but will likely get worse.

Systems Include Mechanisms for Stabilization

If we want to improve the function of a system, we must pay special attention to how the components of the system interact. If our system is a company, we need to understand how the people, the departments, the policies, and so on, work together. Obviously it is not simple. Even when we attempt to improve the right things, we often fail because of the stabilization principle of systems theory: systems include mechanisms for stabilization. Complex systems are stable and in balance. If they are not balanced, they change very rapidly until they reach a balanced state. That balanced state can be desirable if the system operates at a highly effective level, or undesirable if the system functions very poorly. Either way, stable systems can be very resistant to change.

In our daily lives we constantly confront stable systems. A good example is the system that maintains our weight at some level, which for many of us is higher than we would like. A person's weight may drift over a period of years, but it is remarkably stable. The system behaves as though there were a thermostat. In fact, it is just as hard to gain weight above this point as it is to lose weight below the point. In a recent study (Rosenbaum et al. 2003), subjects were asked to either gain or lose weight. The researcher found that for those who lost weight, energy efficiency of the muscles was increased (by some balancing mechanism), meaning that to maintain their weight loss, the subjects had to eat less and exercise more than was required to produce the loss in the first place. The subjects who gained weight became less energy efficient, so that they had to increase eating and reduce exercise still further to maintain the weight gain. These metabolic changes apparently serve as part of a regulatory mechanism that helps the system to maintain a constant weight. In a word, they are the balancing mechanisms that maintain the stability of the system.

In most of the plants we have studied, the safety process is stable. Although accident rates vary from month to month, there is usually no statistical trend. Trends, or their absence, must be determined through the use of control charts, which we will describe further in Chapter 4.

Remember that stability in systems is not necessarily good or bad. Stability is good if the system is producing what you want. It is bad if you are dissatisfied with the output. The stability of one's weight is wonderful for someone at the proper weight. For a person 50 pounds overweight, it is a serious problem. Of course, stability can eventually become a problem, even if it is not one now. Performance in a business that is perfectly good today may be uncompetitive five years from now.

A Stable System Will Not Change Itself

According to Dr. Deming, a stable system in business will not change without introducing profound knowledge from outside the system. The important point here is that without new understanding of the system you want to change, your efforts to change it are likely to be wasted. As Dr. Deming said, "Best intentions are not enough." Without new knowledge, good intentions will only result in wasted effort. The hedgehog concept described by Collins is a powerful example of how new knowledge can change a system. A critical aspect of the new knowledge that changed these companies was an understanding of how to measure success.

WHEN SAFETY IS PUT IN COMPETITION WITH OTHER PRIORITIES, SUCH AS PRODUCTION, A BALANCING SYSTEM IS ESTABLISHED THAT WILL LIMIT BOTH SAFETY AND PRODUCTION.

Attempting to increase safety emphasis is a strategy based on best intentions instead of increased understanding of the system. Increasing emphasis on safety must imply a decrease in emphasis on something else. After all, there is only so much emphasis to go around. Increased safety emphasis is a frequent consequence of a serious accident. When safety is put in competition with other priorities, such as production, a balancing system is established that will limit both safety and production. As safety emphasis increases and the memory of the accident recedes, attention shifts to problems of productivity. Of course, a side effect of this strategy over time is that employees learn that priorities are temporary. This creates cynicism and also makes it increasingly difficult to alter actual practices. The only way out of the impasse is to develop solutions that improve both safety and productivity. Improved work processes offer both advantages. To the extent that we can simplify a process and improve its predictability, we generate improvement in both the safety and the productivity of the process.

Our research (Carder and Ragan 2003) has shown that plants with better productivity have better safety performance as well. Although safety and productivity may compete for emphasis, apparently both respond positively to improvements in the management system.

Obvious Solutions Often Make Things Worse

Any action taken in a complex system has multiple effects. A good example is the use of drugs to treat illness. We say a drug has a *therapeutic effect* and a variety of *side effects*. The only difference is that the side effects are effects that we are not attempting to utilize at the time. If we give morphine for pain, constipation is a side effect. If we give morphine for severe diarrhea, euphoria is a side effect. Sometimes the side effect becomes more important than the original therapeutic effect. Hytrin, a drug originally intended to treat high blood pressure, has the side effect of shrinking the prostate in men afflicted with benign enlargement of the prostate. Hytrin is now prescribed frequently to treat prostate enlargement.

Of course, side effects are not typically beneficial. In fact, they are often so detrimental as to completely outweigh the therapeutic effect of a drug. Thalidomide, a very effective tranquilizer, is perhaps the most widely known, dramatic, and tragic example. Thalidomide produced severe birth defects when administered early in the course of a pregnancy.

By the same token, we should expect any intervention in a business to have multiple side effects. Some of these would be immediate and some would be long term. Because of these side effects, intended solutions often have an effect that is opposite to that desired.

An excellent example of obvious solutions making things worse is safety-based incentives. If employees are paid bonuses not to have accidents, they may pay more attention to safety, in the short term. A common side effect is that they are less inclined to report accidents (Pransky et al. 1999). For a time, things may appear to be better. There may be a reduction in accidents, and there will probably be a reduction in reported accidents. In the long term, some hazards will go undetected, because the accidents they cause will go unreported. As this goes on, employees will become cynical and feel that management cares more about the accident record than about the safety of employees. This will further deteriorate the safety process. Eventually, this may all collapse when one or a series of very serious incidents occurs. Another long-term effect of an incentive plan is that it frequently becomes seen as supplemental income or perks. Even though it may have started as a temporary program, if the plan is removed or changed, it will very likely be seen as a withdrawal from a commitment to safety or as an unwarranted reduction of compensation. This causes tension and distraction that often results in increased accidents.

This is not an argument that the people in the system are usually corrupt and conceal incidents deliberately and flagrantly. When there is a conflict of interests, and when there is a question as to whether something should be reported, there is likely to be an influence on the decision. This may be at an unconscious level, but it is real and measurable. Thus management becomes more interested in how an accident is classified than in what caused it in the first place and how to keep it from happening again. The pressure can be severe and misdirecting. A field manager of a retail chain actually said, "I would rather see a customer hurt than one of my employees. If a customer is hurt it does not go on my record." This approach is counter to the objective of reducing the occurrence of accidents. This manager was focusing on managing the output not on correcting the process itself. This is analogous to putting your finger over the end of a hose to stop the water flow instead of closing the valve. It may work for a while, but eventually you will learn the end-of-line control approach is all wet.

This is not to say that compensation should not be related to safety accomplishments. Because compensation is one way a company makes a statement about what is important, safety should be involved. However, in practice, this has many pitfalls. If compensation is tied to accident rates, then there is a great risk that managers will spend more time trying to manipulate the classification of events. A simple way to reduce recordable accidents is not to record them. In the extreme case, even serious accidents will be concealed. This is not at all infrequent. The solution is not simple. One requirement is that the people who ultimately make these compensation decisions need to have a much better understanding of what is going on. Merely looking at superficial statistics is dangerous, because these are always subject to manipulation. In order to properly determine compensation based on safety accomplishments, a variety of measures should be required. In Chapter 3, we will begin to outline some approaches to this measurement.

A Complex System Has Leverage Points for Change

In order to change a system for the better, we must understand the system and identify a *leverage point*. A leverage point is a point at which efficient and effective action can be taken to initiate change. Action at the leverage point will not create side effects that defeat our purpose. For example, if we decide that we are overweight, eating less is the obvious solution, but it is not a particularly good leverage point. This is probably not news to anyone who has attempted dieting.

The problem lies in the fact that there are mechanisms that seek to regulate one's body weight, in spite of the fact that it is not the weight we prefer. When we diet, these mechanisms are activated. At least two things happen: our appetite increases, and our metabolic rate decreases. Even if we loose weight at first, we are likely to gain it all back. In several studies of individuals who engaged in an intensive weight-loss program with very-low-calorie diets, only 25 percent maintained a 10 percent weight loss after seven years (Anderson et al. 1999). Adding a regimen of moderate exercise to the low-calorie diet improves the results somewhat (Wing 1999), but anyone who has attempted to lose weight knows that we have not yet found the leverage point.

An apparent example of leverage that is of great social importance involves attempts to influence birth rates. Promotion of birth control is typically ineffective. The reasons often lie in the social systems that support high fertility rates. In 1975, Brian Arthur and Geoffrey McNicoll went to Bangladesh on behalf of the Population Council to study that country's population problem. In spite of the free availability of birth control, extensive government programs to control population, and widespread understanding that overpopulation was destructive to the nation, rural families were producing, on average, seven children. Arthur and McNicoll interviewed people in the villages and discovered a rather simple explanation. Because children could go to work at a very early age, they were a benefit to the family. If a woman was widowed, her relatives might come and steal all of her property. Children could defend her against this. Of course, the leverage point is the position of widows in the society, not the availability of birth control or encouragement to use it. Unless the position of the widow changes, even the most draconian measures are unlikely to succeed. Arthur and McNicoll published a paper explaining this (which was promptly banned in Bangladesh) (Waldrop 1992).

Subsequent studies have suggested there are leverage points for reduction of birth rates. There exists a very strong negative correlation between fertility rates and two variables: proportion of women in the workforce and years of education for women (McClamroch 1996). Moreover, it appears that increasing this education level reduces the fertility rate, indicating a causal relationship (Wickrama 2002). Exactly how this might happen is a proper subject for continued research, although many hypotheses have been put forward to explain it. One hypothesis is that increasing the potential wages a woman can command raises the value of the woman's time and thereby increases the cost of childbearing. A second hypothesis is that women with more education tend to marry later and know more about contraception.

This is an excellent example of the principle of leverage, because it emphasizes the indirect nature of many high-leverage solutions. Direct approaches to reducing fertility, such as promoting birth control and creating negative incentives for having children, have generally failed, in spite of the expenditure of considerable resources and significant force. Whenever we wish to alter a complex system, whether it is a living organism or a business organization, we must consider the principal of leverage. Next we will look at an example of leverage in a business.

Several years ago we assessed the safety system in a chemical plant with a fairly good safety record. In looking for instances where procedures were not followed, we discovered that the procedures for mixing ingredients in order to produce a particular chemical would change when the product had to be rushed to the customer. Instead of adding each ingredient in a timed sequence, workers would dump all of the ingredients into the reactor at once and process the batch. In fact, this particular situation did not cause a safety hazard. The workers claimed this did not cause any quality problems either. However, product quality was an issue at the plant. Management was previously unaware of the violations we had discovered.

The obvious solution was to round up the employees and insist that they follow procedures. We explained to management that this would probably not solve the problem. The reason that we did this was that we had looked at the larger system and culture of the company.

The success of the company had been built on its attention to taking care of customers. The operators who were changing the procedure were doing this in order to meet the customer's requirements. A customer had called and announced that if he did not receive the product very soon, he would have to shut down production until the product arrived. Because serving the customer was a very powerful and useful value in the company, we did not want to discourage this or fight against it. In fact, it was probably better for the employees to use improper procedures when the alternative was to force a customer to shut down an operation because of a lack of raw material.

Of course, in a chemical plant or any other plant, it is very important to have predictability and consistency in procedures. Although this particular process did not increase risk, the ad hoc alteration of other processes in the plant might prove extremely dangerous.

Therefore, we looked into the source of the rush situations. A major cause was slowness in paperwork getting through the administrative system. Orders were taking as long as two weeks to reach the production floor, creating rush situations. We suggested working on this as a leverage point. Of course we emphasized that procedures must be followed and provided an explanation of why this was the case. However, we also took action to mitigate emergency calls to the factory from customers who were nearly out of raw material. This was a leverage point in the system.

Two years later, the accident rate of this plant was reduced by 50 percent. The major effort was the involvement of quality teams to improve processes in the plant, such as order entry. This improvement was not the result of a safety awareness campaign. It was attributed to the site's embracing the idea of finding leverage points in processes and following two of Deming's points: *Improve constantly and forever the system of production, planning, or any activity that will improve quality and productivity and thus constantly decrease costs* (point 5). *Put everybody in the company to work to accomplish the transformation. The transformation is everybody's job* (point 14). It was not costly, and the efforts led to significant improvements in quality and productivity as well as safety.

System Archetypes

Although moment-to-moment predictions of output are not possible with complex systems, they often behave according to regular patterns, called *system archetypes*. Understanding these regular patterns, or archetypes, allows us to make predictions and may help us understand what actions to take to improve the system.

Escalation. One of the simplest of the systems archetypes, and one that we have all probably observed and identified, is the escalation scenario. A prominent example can be found in international terrorism. The obvious solution to terrorism is to punish it. Each terrorist act is met with stern punishment or reprisal. Each punishment is seen as unjust by the terrorists and is therefore met by increased terrorist activity. The terrorists believe they are right and that they must retaliate. The targets of terrorism also believe they are right and that they, too, must retaliate. Who is actually right does not have an impact on the outcome. The pattern continues.

In business, price wars are an example of the escalation scenario. Again, each price reduction made by one competitor is met by a new round of price cuts by the others.

Escalation scenarios can persist until the resources of one or both parties are exhausted. They can be ended if one side has sufficient resources to defeat the other. If one side is stronger but not sufficiently strong to defeat the other, then the process can only be ended through the willingness of the stronger party to compromise. Several years ago we consulted with a large company where we found an ongoing escalation scenario between the union and management. We asserted, at a meeting of the company's board of directors, that without ending this, the company could not improve its performance. The company was unwilling or unable to confront the issue. They recently filed a Chapter 11 bankruptcy.

In safety, we can see escalation in the area of discipline. Management discovers that an accident was caused by a violation of safety rules and initiates punishment. Hourly workers perceive that they are breaking rules in order to get the job done, and this is what management really wants. They feel that as long as nothing goes wrong, management ignores rule violations, and that punishment is therefore unjust. Workers stop reporting incidents and conceal true causes of the incidents that are reported, lest they be blamed. Management redoubles its efforts to find wrongdoers and increases the severity of punishment. This cycle can be stopped by either party. In practice, management, the stronger party in terms of control and resources, is the one most likely to be able to end it.

Tragedy of the commons. In an apocryphal story told by Garrett Hardin, the ranchers in a community are allowed to graze their sheep on common land owned by the community. There is no charge for this, so there is an incentive for each individual to graze as many sheep as possible on the public land. However, if too many ranchers take this approach, the commons will be overgrazed and destroyed as a grazing area. Thus each rancher has a longer-term incentive to not overgraze. However, if some ranchers behave according to enlightened self-interest and reduce the number of sheep they graze, other ranchers may take advantage of this to add to their own flocks on the commons. Enlightened self-interest alone will not solve the problem. People who display greed rather than enlightened self-interest will be rewarded in this system. Such a problem can be dealt with only if a higher authority allocates the resource. In a country, a principal function of government is to manage the commons. In a company, this critical function belongs to senior management. The U.S. government allocates such common resources as water, public land, and radio and television broadcast frequencies. Without such regulation, these resources would be depleted, and a few would profit at the expense of the larger community.

In corporations, we can observe the tragedy of the commons scenarios frequently. One of the resources that is frequently overgrazed is managers' time. Although subordinates' time represents a direct cost to line managers, managers' time does not represent a cost to corporate staffs. To the extent that staff organizations can impose requirements on line organization, they are grazing on a common resource. As more and more staff units begin to impose requirements, request reports, and so on, employees' time in the organization is overgrazed, and it becomes difficult to do any single task properly.

This overgrazing by staffs is often a side effect of reorganizing efforts. The reorganizing staffs are under pressure to produce results, and they are given substantial authority over the line operations. We frequently observe middle managers rushing from meeting to meeting so often that it would be impossible for them to have the time to prepare for any one of them properly. This problem can be solved either by refusing to give staffs the authority to make any demands, by a higher level of management regulating the allocation of time, or by establishing some cost for the staff's use of line managers' time.

Safety managers, as part of a staff organization, must be conscious of this issue. To the extent that they can show line managers they are adding value, they will be welcomed. To the extent that they are merely seen as using up time they do not have to pay for, they will meet continuous resistance.

Another example of the tragedy of the commons familiar to those of us in the safety profession is overgrazing of time and resources by regulatory agencies. Regulators can command the time and resources of a business without incurring any cost. If the regulatory agency is judged by the number of citations issued or violations discovered, there will be an incentive for them to use an increasing amount of business resources in order to meet their production goals. This is equally applicable to collecting fines associated with citations and notices of violation. In most regulatory agency charters, agencies are specifically denied the monies collected from fines, to avoid an obvious conflict of interest.

We recall a conversation with the captain of an oil tanker who said that, as a result of the Exxon Valdez incident, the number of overlapping regulators who called on him when he put into a U.S. port was so great that he rarely got much sleep when he was in port. However, he had to be at his sharpest when leaving the port to ensure safety. Several agencies, ranging from the port authority to the fish and game department, were regulating his work. In this way, they could demonstrate to the taxpayers that they were protecting the taxpayers' interest. In fact, by using so much of the captain's time that he could not sleep, they were producing the very opposite effect. They were making an accident more likely.

Fixes that fail. Over the years, many safety professionals have told us that safety programs get stale and that to maintain the safety of their organization, they needed to regularly find fresh approaches. While this may seem logical, we have never seen evidence that this is so. What it sounds like is that these managers are caught in the fixes-that-fail archetype. This relates to our previous discussion of increasing safety emphasis as a way to improve safety.

Increased emphasis on safety puts it into a position to compete with other objectives. When safety emphasis is increased, time and attention will be diverted from other important activities, such as quality, maintenance, planning, and production. As needs increase in these areas, they will demand more attention, and safety will then get less attention. Implementing new emphasis programs simply preserves this cycle. The company's accident record remains in statistical control, with no positive trend. The emphasis programs are a series of fixes that fail. They are short-term solutions that have no lasting impact. They are missing Dr. Deming's admonishment to improve the system constantly and forever. The proper solution is to find ways to improve the processes of planning and production in such a way as to improve safety, quality, and productivity, all at the same time.

The fixes-that-fail scenario is often easy to identify in organizations, for example, when employees talk about a flavor of the month, referring to the fact that a succession of initiatives have come and gone. The cynical tone of the phrase "flavor of the month" implies that each of these initiatives have failed.

There is a short but powerful chapter in *Built to Last* (Collins and Porras 1994) entitled, "No 'Tyranny of the OR' (Embrace the 'Genius of the AND')." Their point was that there is no reason to limit actions for improvement to one business function *or* another. The actions should improve the target business function *and* others as well. This is the idea of win-win. It is interesting that this fundamental point can be so well made so many times by so many people and still be so widely ignored in business in general and in accident-prevention efforts in particular.

The safety function has a unique opportunity to lead the change process to create a win-win solution, so that the improvement in safety performance is accompanied by improvements in quality and productivity as well. At the risk of using a cliché, if not us, who? If not now, when? We are talking about protecting lives, property, communities, and the businesses that employ us. How can we stop the cycle of failure? It is our hope that every time readers of this book investigate yet another accident similar to one that has occurred other times within their organization or community of knowledge, they consider this challenge and strive to find fixes that improve constantly and forever the system, not fixes destined to fail.

Shifting the burden. Shifting the burden is a common method of problem solving in corporate America. In fact, it may be a common method of problem solving in all human groupings. It consists of taking a problem and making it someone else's problem, in spite of the fact that the root causes of the problem have nothing to do whatever with the party to whom the burden has been shifted. For example, we have encountered a number of situations in which the manager of a unit has told us that, in order to improve safety, employees need to become more aware and more involved in the safety process. These managers tell us that if employees would just take more responsibility, many accidents would be eliminated.

In most cases our analysis shows that what is lacking is involvement of the manager. Although excellent safety performance certainly requires the involvement and commitment of employees, that involvement is strongly influenced by the employees' perception of management's involvement and commitment. What these managers are doing is attempting to shift the burden by making safety the problem of the employees. Correctly viewed, the problem is systemic. Both the employees and management are part of that system. It is unlikely that either group, working in isolation, could do very much. However, management has much more control over the system and is in the best position to institute change regardless of where they try to shift the burden.

Blame fixing is a form of shifting the burden. Identifying any one individual or group as the sole source of a problem is likely to be a case of shifting the burden. For example, let us say that in the Challenger disaster, the real problem was management. Senior managers ignored clear signs of a potential disaster, probably in an attempt to maintain the prestige and funding that the National Aeronautics and Space Administration (NASA) enjoyed by getting positive publicity for meeting a scheduled launch. However, it is not that simple. Congress, which makes funding decisions on the basis of current publicity, certainly added to the problem. Of course, Congress is elected by the public. Then there were the NASA engineers, who had good reason to believe that a disaster was imminent but did not take the matter far enough to prevent it. What about the rocket-system designers? The point is that looking at the problem as systemic usually reveals blame fixing as a silly exercise. Rather than blaming people, we need to fix the system.

It was reported that employees at NASA during the period before the Challenger accident often quipped "we the few, we the tired," as a jab at the understaffing of the program. At first glance this would seem to place blame on management for pushing the program so hard and Congress for underfunding the program. But, as noted above, when you see blame being pushed, the ones pushing frequently are doing so because they recognize they own part of that blame but wish to push it off on someone else to avoid the burden, work, or responsibility. In this case, as with others discussed in this book, there is always enough blame to go around. Until we recognize that we must change the system rather than blaming the components, we will not get the results we want.

SCIENCE AND MANAGEMENT

In spite of the fact that business is a strong supporter of science and technology, the science that applies to the management and leadership of the business itself is often ignored. Dr. Deming's innovation might be viewed at the simplest level as a reasoned and thorough approach to applying existing science to the management of organizations. Although managers will not often know any of the disciplines of profound knowledge at a professional level, they must be sufficiently aware of this knowledge to avail themselves of professional advice on the frequent occasions when this knowledge is critical to the manager's decision process.

KEY POINTS

This chapter described what Deming called profound knowledge, essentially the particular branches of scientific knowledge applicable to leadership and management. The four branches are:

- **Variation:** Understanding variation is essential to drawing the proper conclusions from numerical data. Absent this understanding, unwarranted actions are frequently taken. These actions increase cost and reduce safety and quality.
- **Psychology:** Most managers and all leaders need to work effectively with people and hence need an understanding of psychology. Much of what passes as common knowledge about psychology is wrong. This leads to an overuse and improper use of incentives and a failure to understand intrinsic motivation.
- **Theory of knowledge:** This sets standards for how knowledge is created and emphasizes the importance of theory. According to Deming, there is no knowledge without theory. An important application of this is that employees can do a better and safer job if they understand some of the theory behind their work.
- **Theory of systems:** A business is a complex system. Most systems are stable and resistant to change. Understanding the theory of systems can assist us in diagnosing system problems and in crafting leveraged solutions that positively impact the system without causing undesirable side effects.

ENDNOTES

1. The recordable accident rate is the number of recordable accidents per 200,000 hours worked. This approximates the total hours that would be worked by 100 employees in a year. When hours of work are not available, the rate is computed as the number of recordable accidents per 100 employees.
2. Risk = hazard x probability.

3

Some Principles of Measurement

Before you initiate a trip to a new destination, it is a good idea to obtain a map of the territory. An effort to improve a business process is, in essence, a trip to a new destination. Measurement can provide us with a map.

THE PLAN-DO-STUDY-ACT CYCLE

Deming explained that the process for improving a system is the cycle of *plan-do-study-act* depicted in Figure 3.1. He credits this formula for process improvement to his mentor, Walter Shewhart, and calls it the Shewhart cycle. Others have called it the Deming cycle or the PDSA cycle.

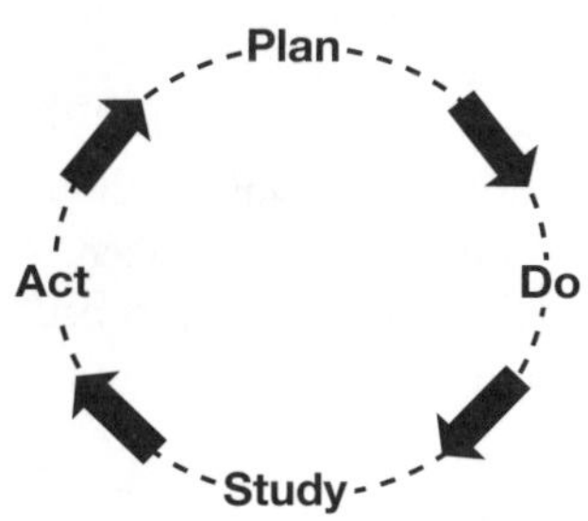

Figure 3.1 The plan-do-study-act cycle.

It is the fundamental method of Deming's system of quality management for the improvement of a stable system. It begins with the development of a *plan*. Planning involves collecting and analyzing data about the system. Knowing where to start is often a difficult problem. An important source of data is observations made by people who work within the system. There should also be other sources of data about the system. For example, if you want to increase the flow of product through a particular stage of production, it would be useful to know employees' opinions of obstacles to production. It would also be necessary to know the rate of product flow through each step, along with the variation in that flow. Realize that asking for employee observations and charting the flow of product through the system are both examples of measurement. Of course, there are many other measurements that may be necessary to provide you with sufficient information to develop an effective plan. Based on the information you assemble, you determine the obstacles to improved performance and identify potential leverage points in the system. The action plan is designed to act on these leverage points to improve the system.

THE ABILITY TO ANALYZE DATA AND DERIVE USEFUL CONCLUSIONS IS A PROFESSIONAL SKILL.

It is our experience that making effective use of the information is actually more difficult than getting the right information. The ability to analyze data and derive useful conclusions is a professional skill. Deming frequently spoke of the fact that businesses collected enormous quantities of data in their computer systems but made little use of it. His exhortation was to get the data off the disk. He strongly advocated plotting charts by hand with paper and pencil. The two of us no longer do it that way, but we do make many of the computations ourselves, rather than order up a set of graphs. The analysis of data is best done with as much direct involvement with the data as possible. Although the principles are scientific, the process is an art, developed through experience.

How much data you need to collect is another difficult question. Hopefully, as you collect and analyze information, you will begin to develop hypotheses about what actions should be taken. At this point, you need to develop a plan. Your analysis may have positioned you to have extreme confidence in the plan, or you may be at the level of exploring hypotheses. If you have not developed even a reasonable hypothesis, you need to continue looking at information. If this has gone on for a long time, and if you are working on a very important problem, you should seek help from someone more experienced in this process. It is important not to act unless your planning phase has provided direction. If you do have direction, you proceed to the *do* phase. In the do phase, the plan is implemented, on a small scale if possible.

The next phase of the process is the *study* phase, where the results of the implementation in the do phase are studied. Again, several data sources should be available. Much of your measurement would actually be determined in the planning phase, where you have laid out your objectives. If you want to increase product flow through a particular stage of a process, you would obviously measure that flow. You might also ask the operators of that process to describe the effects of the changes you have made, both positive and negative. As with the planning phase, measurement should be carried out with adequate knowledge of statistics and scientific method. Based on the results, the plan may be revised.

When you complete your study, you can proceed to the *act* phase. In this phase, plans found to be effective are implemented on a wider scale. The cycle repeats again and again to generate continuous improvement. If the study phase shows that your plan is not effective or that it needs modification, you return to the plan phase.

When we began to apply this process to safety, we already had extensive experience with successful application of the PDSA cycle for improving manufacturing and service processes. It was logical that the model would work for safety improvement. The first challenge is always measurement. Planning requires information about the system. To the extent that you can develop information that helps you understand how the system works, your plan will be successful. It is not how much data you have but whether you have measured the proper thing. This is not trivial. All too often businesses measure what is easy or what is obvious.

This leads us to another dangerous oversimplification: what gets measured is what gets done. This theory is widely believed by managers and is likely the truth. To be more specific, managers will usually attempt to produce the results by which their superiors judge them. If accident statistics are the key measure of safety performance, the managers will focus on reducing incidents. Management based on this theory is a relatively simple task: set some requirements based on a clear set of measurable objectives and hold your subordinates accountable to meet their targets. If they do, they get a bonus. If not, they are reprimanded or terminated.

After a combined 60 years of managing and observing the operation of business, we know it is not quite so simple. Part, but not all, of the problem lies in the two corollaries that we would add:

- What gets measured is often not the correct thing.
- Therefore, what gets done is often not the correct thing.

WHAT GETS MEASURED IS WHAT GETS DONE:

- WHAT GETS MEASURED IS OFTEN NOT THE CORRECT THING.
- THEREFORE WHAT GETS DONE IS OFTEN NOT THE CORRECT THING.

A straightforward example of measuring the right thing is seen in a project that we conducted many years ago. In 1988, one of us was engaged by a small company in Silicon Valley to help implement Deming's methods. This company made disks for hard drives.

Interestingly, the highest technology in hard drives was the manufacture of the disks that stored the information and the read/write heads that read and wrote the information on the disks. Most of the drives were being assembled in Singapore, but the disks and heads were made in the United States. At that time the maximum capacity of a laptop drive was about 20 megabytes. Moore's law, based on an observation of Gordon Moore, one of Intel's founders, was that data density would double every 18 months. We knew that our manufacturing quality had to improve rapidly. Today, 14 years later, drives of 60 gigabytes are not uncommon. This 3,000-fold increase over that period is about what Moore's law would predict. This increased capacity requires a dramatic increase in the uniformity of the disk itself. Each bit of memory has to be stored in an ever-smaller space and read by a read/write head that is closer and closer to the surface. Our job then was to help the company improve the uniformity of the product as rapidly as possible, both to meet the need of current customers and to enable increased storage capacity that would be required in future products.

The disks the company was making were coated with a thin film of material for the magnetic storage of information. The film was applied to the disk by a process called *sputtering*. The parts to be coated were placed in a high vacuum. A block of the coating material, called the *target*, was vaporized by a stream of high-energy electrons. The coating material was then deposited onto the disks. This method is capable of creating a uniform and very thin coating. It is used to deposit conductors on integrated circuits in the manufacture of computer chips.

The company's sputtering machines were about the size of a Chevy Suburban car and cost in the neighborhood of one million dollars each. The machines had air locks that permitted nearly continuous production. The guaranteed interruption came when the target material was used up. Every three to four days, a new target had to be installed. This was no small task, for one important reason. The vacuum level required for production was so great that it took as much as 20 hours to pump the machine down to the required vacuum level.

The theoretical minimum time for a target change was thought to be about 24 hours from the cease of production to the restart of production. This included 4–6 hours to disassemble and reassemble the machine and 16–20 hours required to achieve a production level of vacuum in the machine. The company was averaging 36 hours per target change, which was a serious problem. Since the company could sell every disk they made at that time, the lost production time was costing them millions of dollars in revenue over the course of a year. We assembled a team of engineers to address the problem.

The 36-hour average was interpreted as indicating that the disassembly and reassembly process was too slow. The engineers quickly offered a solution. The problem, they said, was the machines. In order to change the targets, several doors had to be removed. These doors were held in place by numerous screws. The problem was the design of the doors and the screws. The engineers had an idea for a new door that would be much easier to remove and replace.

Were it not for the cost of this redesign, we might have gone along with them. Based on what we had learned about equipment costs, we guessed that the modifications they were suggesting would run in the hundreds of thousands of dollars per machine. Therefore, we suggested we look into this a bit further before implementing this logical *engineering* solution.

We asked for a chart of the target change times over the past few months. What was brought to us looked like the chart in Figure 3.2. It depicts the times of each target change, in sequence, over the preceding months. Although the average is 36 hours, the chart is characterized by many times in the 24–32-hour range and many in the 40-hours-plus range. The chart showed a bimodal distribution with a mean of 36. The engineers were responding to the mean but ignoring the shape of the distribution. It seemed logical to find out why the distribution was bimodal.

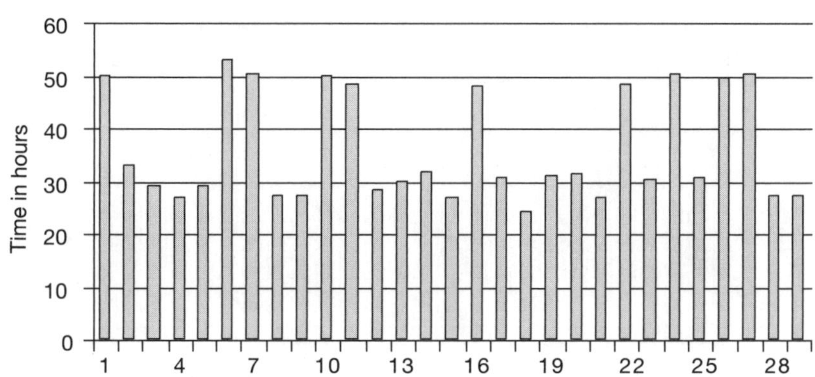

Figure 3.2 Target change times.

We called in the technicians who actually did the work. What was happening, they said, was that sometimes a machine would be opened and the targets replaced. The machine would be closed and the vacuum pumps started. Remember that the pumping time required to reach a production vacuum level was as much as 20 hours. At some stage, they said, well into the pumping cycle, they would discover that the machine would not hold the necessary vacuum. This would turn out to have resulted from an error in the reassembly process, such as a washer omitted, a screw that was not properly torqued, and so on. They would have to stop and repeat the disassembly and reassembly process.

When we asked why this was happening, we were told that the reassembly had been done improperly the first time. They went on to say that this frequently happened when the job of disassembly and reassembly was started on one shift and passed to the next shift before completion. The communication between the initial shift and the second shift had broken down somewhere.

At this point, we could have focused on this communication process. However there was a much simpler solution. The disassembly and reassembly took about 4–6 hours. Shifts were 10 hours. We made a rule that the entire process of disassembly and reassembly must be completed on the same shift. A supervisor who saw that a target would be exhausted toward the end of the shift would have to order the target change to begin earlier so that the assembly process would not be split between two shifts.

The result of implementing this rule is depicted in Figure 3.3.

The bimodal pattern was gone. Although we were not at the theoretical limit of 24 hours, we had accomplished a 25 percent improvement at a minimal cost. We had very slightly increased the frequency of target changes. Of course our change had benefits beyond the additional production time

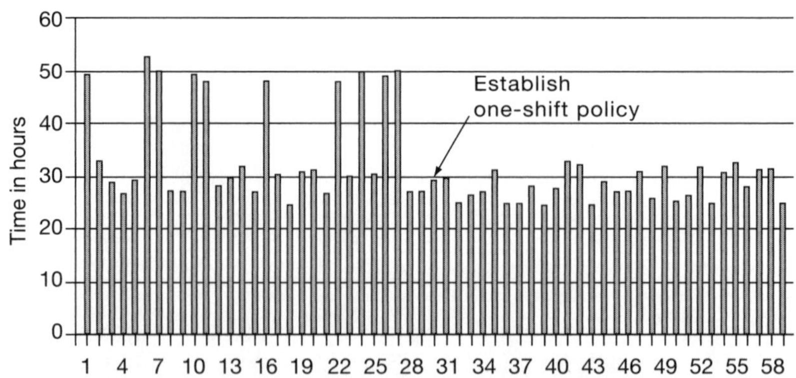

Figure 3.3 Target change times under the one-shift policy.

gained. Any time you improve the uniformity of a manufacturing process, you will likely improve the uniformity of the product. This is the essence of quality as Deming explained it. In addition, with the greater predictability of the process, the engineers were actually able to initiate preventive maintenance and cleaning during the target change process, again improving the quality of the product.

The critical step to solving this problem was to look at a simple plot of the raw data so that we could see the distribution of the times in addition to the mean. Focusing on the mean also would have led to solutions that amounted to speeding up the process or purchasing expensive technology. By looking at the distribution, discovering that it was bimodal, and then asking the workers why this was so, we arrived at a solution that was immediate and cost virtually nothing. Once we understood what the problem was, the solution was easy. Had we merely focused on the mean, we likely would have never found this simple solution.

In Chapter 1 we discussed Collins's book *Good to Great*. We described the hedgehog concept, which was one of the two fundamental steps in transformation to a great company. Part of this concept was the determination of what the company needed to measure. For example, it was critical to Walgreens to measure profit per customer visit, rather than profit per store or profit per territory. This was not a simple or trivial decision. Collins reported that companies took an average of three years to develop a successful hedgehog concept.

A brilliant example of measuring the right thing is reported in the recent book *Moneyball* (Lewis 2003). The book describes how the Oakland Athletics, with a payroll that was 25th out of 30 teams in the major leagues in 2002, manage to win year after year. The Oakland system is based on measuring the right thing. The objective is to select a team of players who will win games but whose salaries will stay within Oakland's modest budget. The first task is to understand what players do to win games. The answer, for position players (nonpitchers), is that producing the most runs will win games. The next question is what statistics of individual players predict run production. The most important factors are on-base percentage and slugging percentage,[1] with on-base percentage being three times as important as slugging. On-base percentage is equal to hits, plus walks, plus being hit by the pitch, divided by total at bats. The league average is about .340. Baseball professionals value a high batting average more than on-base percentage. Every year there is a title for the highest batting average, but there has never been a title for the highest on-base percentage. A hitter who hits .310 but rarely walks and has an on-base percentage of .330 is not someone that Oakland would sign. His batting average would demand a high salary, but his on-base percentage would tell Oakland that he is not a great producer of runs. What Oakland looks for is players with statistics that predict run creation but are undervalued in the marketplace. A player with a low batting average who gets a lot of walks is ideal. Whereas players with high batting averages are highly paid, players with high on-base percentages are paid much less if they have a low batting average. In addition, players who can run fast are paid disproportionately to their contribution to run production and therefore to winning. It turns out that stealing bases is a very risky proposition. One of the worst things a team can do is have a base runner make an out. If the success of a base stealer is under about 66 percent, then his stealing attempts are actually costing the team runs. Because the contribution of speed is overvalued, Oakland has a slow team. By focusing on a different set of measures than their competitors, Oakland was able to gain huge economic leverage and compete effectively with teams like the Yankees, who have nearly three times the payroll. Note that these measures were not lucky guesses. They were measures that were carefully validated by statistical analysis.

In safety almost everyone looks at incidents. They are observed in a variety of ways. We measure frequency, severity, and cost. We try, sometimes with success, to understand the causes. Most companies with significant risks also use audits. Again, these are done in a variety of ways. Less frequently, companies observe and record the behavior of workers, typically classifying the observed behavior into safe and at-risk categories. Finally, there are occasional attempts to measure the management system or the organizational culture, using written surveys and interviews.

All of these measures presumably have the same objective, namely, to provide information that will enable the prevention of future accidents. Using the PDSA approach to safety improvement, measurement is critical, because without good measurement neither planning nor study is even possible.

The question is, what is *good measurement*? This is not just a matter of opinion. There exists a science of measurement that must be applied here.

THE SCIENCE OF MEASUREMENT

THE QUALITY OF A MEASURE IS DETERMINED BY ITS RELIABILITY AND VALIDITY.

The quality of a measure is determined by its reliability and validity. No measure is perfect, and all measures have limitations on their reliability and validity.

Reliability

Essentially, reliability refers to the repeatability of a measure. For example, if we want to measure the width of a room, we could pace it off, use a tape measure, or employ a laser. If we measured the room 10 times with each, working at the limits of each method, we would likely get a spread of numbers around an average in each case. We call this spread variation. We quantify variation by computing the standard deviation. The variation of measurements from the laser, presuming it was working properly, would be much smaller than the others. The pacing would have the most variation. We would conclude that the laser was more reliable. There is no such thing as a perfectly reliable measure. All measures will show variation. Deming pointed out that there is no true value of anything. There is a measurement method and a result. Using different methods usually leads to somewhat different results. His example was the extreme variation in measurements of the speed of light, depending on the particular method used.

In some cases we assess reliability by looking at the variation between observers. For example, audits are scored on the basis of the judgment of an auditor or audit team. To judge the reliability, we would ask whether a different audit team, unaware of the initial team's evaluation, would give the same or similar score. In practice we would have two auditors conduct audits of a number of sites. Each auditor would be unaware of the scores given by the other. We would then compute a correlation coefficient between the scores given by the two auditors. Analyze the results, and remember that the larger the coefficient, the greater the reliability. A very low or even negative correlation would mean that the process has no reliability at all in this particular test. It is still possible that one of the auditors is very accurate. However, there are two problems with this: (1) we do not know which one was correct, and (2) even if we could find out, our process would be dependent on one person's judgment. In this case our measure is neither reliable nor useful in a practical sense. In order to develop a reliable and useful measure, we would attempt to clarify and better define the criteria and methods, retrain the auditors, and test again for reliability on a different set of plants.

No matter how reliable a measure might be, it will still have variation. If two auditors consistently report exactly the same score for the audit of a plant, we should question whether they are really operating independently. In fact, anomalies in variation often signal a problem with our measurement process. If a measure does not have any variation at all, we are not looking closely enough, the gage is broken, or the numbers are being manipulated by the observer. It is important to understand that numbers are often invented or fudged by the person reporting them.

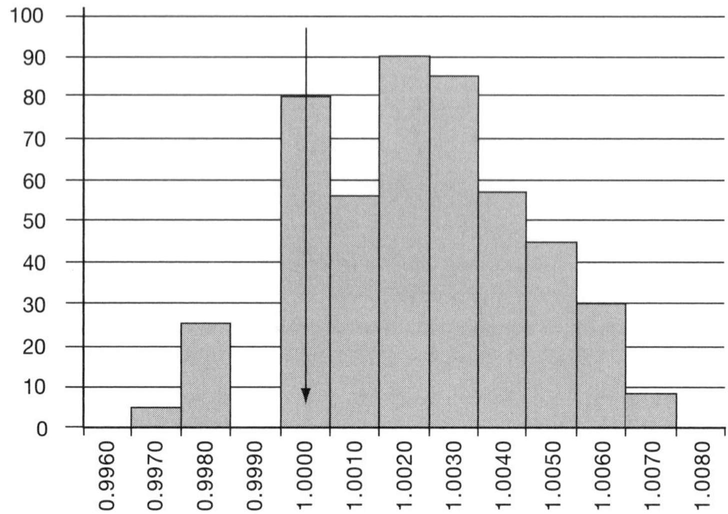

Figure 3.4 Manipulated measurement of steel rods.

The manipulation of numbers is most often revealed by observing anomalies in the variation of the measurement. Usually, in the case of fudged numbers, there is some variation, but the pattern of variation is not appropriate. There are countless examples. Figure 3.4 is taken from Deming's *Out of the Crisis* (1986). It represents the measurement of the diameters of 500 steel rods as made by an inspector. Rods with a diameter of less than 1.000 had to be rejected. The distribution should have been a normal bell curve, symmetrical on both sides of the mean. What was happening was that the inspector was passing rods that were a little too small, hence the lack of any parts measuring .999 and too many rods at 1.000.

It is very unlikely that this distribution represents the rods produced by the machine. The explanation has to lie in the behavior of the inspector. If too many rods are rejected, people might lose their job, or worse yet, the plant might close. Surely there will be no problem, thinks the inspector, if we list this .999 rod at 1.000.

RELIABILITY IS A PROPERTY OF THE ENTIRE MEASUREMENT PROCESS: THE INSTRUMENTS, THE PROCEDURES, THE ENVIRONMENT, THE PEOPLE, AND SO ON.

Reliability is not a property of the instruments, but of the entire measurement process. This includes the tools, the instruments, the procedures, and the people. Subjective judgment can be very reliable in some cases, whereas measurement with the finest instruments can be unreliable if the process that uses these instruments is flawed.

As we said then, any measure will vary. If the measure does not vary, or if the pattern of variation is not normal, we need to investigate. Assuming the measure does vary in the appropriate way, then the less the variation, the higher the reliability. For any particular purpose, there is a level of reliability that is acceptable for the task. If we are carpeting a room, we can use a tape measure but not the pacing method. If we are measuring length in order to construct a complex optical system, the laser might be required.

The level of reliability of a measure limits the kinds of decisions that should be based on that measure. A good case in point is the financial statements of public companies. There has been considerable recent publicity about the lack of reliability of these statements. Enron statements showed dazzling profits as the company was plunging into bankruptcy. Based on the statements, people and institutions invested large amounts of money in Enron stock.

Although Enron is an extreme case, there is considerable limitation on the reliability of financial statements even from ethical companies. If we assign two independent teams of accountants to a company, it is likely that they would produce significantly different numbers.

Forbes used to run a contest for accountants. Contestants would all receive the same data from a fictitious taxpayer and would be asked to create a return. There was always a huge variation in the results. And this was a much simpler task than the preparation of financial statements for a large company.

The problem is more complex, however, than the low reliability of accounting measures. Figure 3.5 shows 813 consecutive earnings reports from public companies, as listed on the Bloomberg website. What is plotted is the difference between the report and the earnings forecast. A minus number means they fell short. The expected normal distribution is the black line. A statistical test tells us that the actual distribution is not normal. What is happening is that far too many companies are exactly hitting the target. This is not a case of extreme accuracy of prediction. No matter how accurate the predictions, the distribution should be normal, but with a smaller standard deviation. Here we have a distribution that is not normal, with far too many exact hits on the prediction when taking into account the magnitude of the misses on either side. This could occur because the resolution of the measuring process is not sufficient to show the normal variation. This might happen if we were to measure newly produced pencils with a yardstick. They would all appear to be the same length, because the yardstick cannot make sufficiently fine discriminations of length. That is not what is happening here. Rather, there is a problem

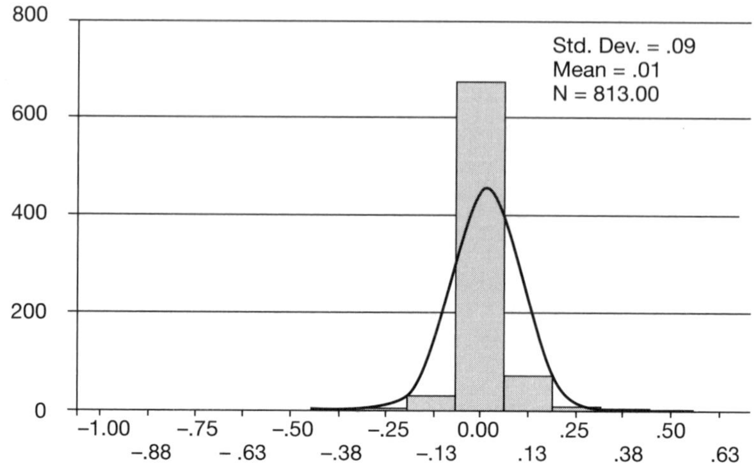

Figure 3.5 Earnings reports from public companies.

with the measurement process that, in this case, probably involves manipulation. Incidentally, manipulation may be to increase or decrease earnings. A company that is about to overshoot its estimate may want to save some of the earnings for the next quarter, lest they fall short or their next target be set at a higher level than they would like. Which companies are manipulating their earnings is not possible to say from this analysis. However, the graph and the statistical analysis strongly suggest that the measurement process that produced these numbers is flawed. We should also point out that this is not necessarily evidence of fraud or malfeasance. Generally accepted accounting principles (GAAP) allow fairly wide latitude in how financial statements are constructed. Working within these guidelines, a good accountant can steer the company in the direction of the target.

Safety data are subject to the same kind of manipulation. We will discuss this in some detail when we discuss incident-based measures in Chapter 4.

In safety, we must realize that all of our measures have limitations on their reliability, and that all measures are subject to manipulation. It is important that we understand the limitations of each measure and we take these limitations into account when making decisions based on the measurements. In the following chapters we will discuss the reliability of a variety of measures of safety.

It is important to realize, also, that a measure with high reliability may still be worthless, because it does not tell us anything useful about the thing we are trying to measure. Just because a measure is reliable does not mean it will help us take effective action. This brings us to the question of validity.

Validity

Validity relates to whether you are measuring what you want to measure. When we measure the width of a room, the question of validity usually does not arise. When we are measuring a complex process, such as aptitude to perform well in college or the ability of the safety-management system in a plant to prevent future loss, validity becomes a serious question. Scientists (Chronbach and Meeh 1955) generally define three categories of validity: content-related, criterion-related, and construct-related validity.

Content-related validity. This is often called face validity. It addresses whether the content of the measurement process is, on its face, related to the purpose of the measurement. A good example is found in safety audits. If one of your questions asks whether employees use personal protective equipment on the factory floor, the question has face validity. If you ask whether the workers go out for beer together after work, that lacks face validity, because nothing in the content appears to have anything to do with safety. However, it might turn out to have the other two kinds of validity. It might turn out that the plant is safer when the employees have close personal relationships, and that going out for a beer after work is evidence of such relationships. By the way, we are not asserting that this is true and are not selling happy-hour specials as a safety program.

Criterion-related validity. This is sometimes called predictive validity. It deals with whether our measure correlates with other measures that could be called criteria. For example, the SAT test is an attempt to measure the likelihood that a student will succeed in college. Obviously the criterion here is college performance. Yale University has used the SAT for many years. Although they admit it is not a very good predictor, they also say that it is the best they have. Over the years, the Pearson correlation between SAT scores and Yale grades has run in the range of 0.2–.03. This is statistically significant, meaning that the SAT does indeed have criterion-related validity. However, this level of correlation means that at best, what is measured by the SAT is accounting for no more than 9 percent of the variation in college grades. (Squaring the correlation coefficient of 0.3 gives us the percent of variation accounted for, 0.09.) The other 91 percent is presumably accounted for by other things, such as motivation, the quality of the student's secondary education, the difficulty of the courses chosen at Yale, luck, and so on.

When dealing with large populations, it may make good economic sense to use measures with relatively low criterion-related validity. However, individuals who are negatively affected by such measures will always have a pretty good argument that the measurement was unfair to them.

It is also important to realize that the criterion is somewhat arbitrary. After all, the success of a Yale career should not be measured by grades. Yale is interested in producing good and productive citizens and leaders. Does the SAT predict that? If that study has been done, we cannot find it.

ULTIMATELY, YOU WOULD LIKE TO MEASURE THE ABILITY OF THE SAFETY-MANAGEMENT SYSTEM TO PREVENT FUTURE LOSS.

In safety, injuries and monetary losses are certainly useful as a criterion against which to test other measurements. However, they are not the only possible criterion, and they may not be the best one. Ultimately, you would like to measure the ability of the safety-management system to prevent future loss. Although a burned finger may be of some concern in a chemical plant, it is trivial compared to a release of a toxic chemical that could injure or kill thousands of people. Because catastrophes are fortunately infrequent, they are inconvenient to use as a criterion in a validation study. An excellent safety measure would enable the prevention of catastrophic events. Whereas the burned fingers are associated with cost and suffering, they are neither equivalent nor directly related to catastrophic events. Minor injuries should not be the total focus of a safety effort. However, because the performance of managers is often based on incident rates, and because most incidents are relatively minor, managers tend to focus disproportionately on these minor incidents.

Because criteria are usually somewhat arbitrary, and because there is often no single, ultimate criterion, it is best to use several criteria when attempting to establish criterion-based validity. Doing so will not only make it somewhat more likely that you will establish this validity but likely increase your understanding of the measure you are testing. This brings us to the concept of construct-related validity.

Construct-related validity. This goes to the understanding of what you are measuring. All of us have probably taken an IQ test in some form. What is intelligence? Most of us have some general concept of what it is. It is how smart you are. It is how fast you can learn. It is how good you are at problem solving. Initially, IQ scores were validated against such criteria as academic performance and economic success. Obviously the correlation was not perfect. The former chairman of the psychology department at an Ivy League university was given an IQ test as a young child. Based on his score, he was assigned to a class for the developmentally impaired. The test was in English. At the time, he could speak only German.

There is an ongoing debate about the concept of intelligence. Jensen (1987) has argued with great influence that intelligence tests measure a general intelligence or G factor but has noted (Kranzler and Jensen 1991) that this appears to derive from at least four separate cognitive abilities, such as speed of visual search and speed of memory search.

Howard Gardner (2000) has identified eight components of intelligence:

> *I have concluded that all human beings possess at least eight intelligences: linguistic and logical-mathematical (the two most prized in school and the ones central to success on standard intelligence tests), musical, spatial, bodily-kinesthetic, naturalist, interpersonal, and intrapersonal.*

By identifying a number of dimensions of intelligence, Gardner was reducing the likelihood of potentially gifted persons being excluded from the educational system just because they had a low score on a test that measured only one or two dimensions in which they did not happen to excel. Important decisions are often made on the basis of intelligence test scores. Scores on the SAT (which Gardner notes is a thinly disguised intelligence test) have a large influence on who gets into what college. It is likely that it has some validity in terms of predicting success in college. Does it predict who will be a good leader? Does it predict who will be a moral person? Does it predict who will stand up in a time of crisis and serve the common good? The point is not so much that we should abandon the SAT but that we should realize the limitations of its construct-related validity.

In the measurement of safety, there has been very little work on construct validity. Safety professionals have tended to take the measures they are using for granted as a result of their content-based validity. They look to the content of their measurements to determine exactly what is being measured. There is a serious limitation in this. The most obvious case is the way we count incidents. A minor cut or burn is recorded and investigated. Yet many more important events, such as a chemical reaction going temporarily out of control, are often neither recorded nor investigated. The assumption, based on faith rather than evidence, is that the burn and the control of the chemical process are the same thing. Although they may be related, we will discuss evidence that they are not the same thing.

A MEASURE WITH MODERATE RELIABILITY AND HIGH CONSTRUCT-RELATED VALIDITY IS MUCH TO BE PREFERRED OVER A MEASURE WITH VERY HIGH RELIABILITY AND LITTLE OR NO EVIDENCE OF CONSTRUCT-RELATED VALIDITY.

As we better understand what we are measuring, we can better judge what management actions are suggested by our measurement. One of the problems we frequently encounter in business is a tendency for managers to prefer measures they believe to be highly reliable, in spite of the lack of any evidence regarding the validity of those measures. Many managers believe that surveys and interviews are of doubtful reliability and that the measurement of incident rates is quite reliable. In fact the data show that surveys and interviews can be very reliable, whereas recording incidents is frequently unreliable. More important is the limited ability of incident rates as a measure to enable the prevention of future catastrophic loss. This is an important limitation of the construct-based validity of incident-rate measures. A measure with moderate reliability and high construct-related validity is much preferred over a measure with very high reliability and little or no evidence of construct-related validity.

The following chapters will examine the reliability and validity of some of the available measures of safety. The objective is not to eliminate some and choose the best but to better understand the strengths, weaknesses, and best use of each.

KEY POINTS

- Improvement plans should be developed using the Shewhart cycle of plan-do-study-act.
- The effectiveness of the planning phase is based in large part on the quality of measurements you have made of the current system.
- The quality of a measurement is based on its reliability and validity. Reliability relates to whether you get the same result if you measure the same thing again. Validity relates to whether you are measuring what you want to measure. No measure is perfectly reliable, and no measure of a complex system is perfectly valid.
- To be useful a measure must be both reliable and valid.

ENDNOTE

1. Slugging percentage is equal to total bases divided by total at bats.

4

Incident-Based Measures

The bulk of the safety-measurement effort is spent on incident-based measures. Essentially this is the recording of accidents and the investigation of their causes. Since 1970 OSHA has required companies with 11 or more employees to maintain a record of accidents and injuries. The purposes of this recording, according to the current OSHA website, are "to permit survey material to be compiled, to help define high hazard industries, and to inform employees of the status of their employer's record." Some companies also record and investigate what Heinrich called near misses, events that might have caused injury but fortuitously did not and therefore did not require recording under OSHA regulations.

There is a general assumption that incident-based measures are concrete and therefore can be measured very reliably. Although there are frequent complaints that incident-based measures are not very helpful in process improvement, there is rarely a question about their reliability or validity. In fact, examining the reliability and validity of these measures will go a long way toward helping us explain the limitations of incident-based measures as a source of direction for process-improvement efforts.

RELIABILITY

There are two sources of limitation on the reliability of incident rates: variation in interpretation of the criteria for recording an incident and variation in the statistic itself. We will address them separately.

Variation in Recording Criteria

In spite of best efforts to the contrary, recording criteria are ultimately quite subjective and vary from company to company, site to site, supervisor to supervisor, individual to individual, and day to day. As an example, the shift in the average incident rate for the chemical industry, and many other industries for that matter, in the late 1980s was more a result of a change in the OSHA recording criteria than of a decline in the effectiveness of the industries' safety-management systems. The criteria were changed in 1986 to allow OSHA to asses a much-publicized highest fine in OSHA's history, primarily because of the accident record-keeping practices of a plant in West Virginia. OSHA changed the rules for this and every other site when they said that other systems, such as the older ANSI system or individually defined systems, could not be used for injury record keeping. OSHA decided to cite each individual log entry that was deemed to be in violation for the $10,000 maximum fine. The traditional practice was to assess a single fine for poor record keeping. This new approach caused many companies to change their criteria and become extremely conservative in their recording of accidents and illnesses. Rather than risk a large fine, they would record any event that even smelled like an OSHA recordable accident.

OSHA changed its requirements again in 2001 resulting in yet another shift in the way companies kept records. One revision changed the requirement for recording hearing losses and led to a temporary upswing in recorded hearing losses which increased recordable rates. Another major change eliminated recording illnesses as a separate class of incidents. Illnesses are now recorded according to the type of treatment given, as opposed to all illnesses being recordable, which should reduce the total recordable rate. Before these changes were made, the quality of the accident-prevention system was probably improving while we observed a flat or rising trend in incident rates. These changes in the measurement methods mean that the changes in incident rates may not accurately reflect changes in accident prevention. Thus we cannot compare incident rates of 1972 to those of 1998 or the 1998 rates to those of today. The intent of these changes represents a commendable effort at continuous improvement. Very likely each of the changes did result in improved measures, but they also resulted in an inability to compare results from year to year.

The fine in West Virginia changed the chief rationale for maintaining accident records from one of identifying trends and measuring management performance to one of satisfying an OSHA requirement. Their function as a management measure moved to a subordinate role. After the $1.27 million dollar fine of Union Carbide, companies began following the letter of the law a lot more closely than they had before. It has been argued this was done to make the numbers more consistent from site to site and company to company. This did not happen. It just widened the gap between companies that monitored and followed the rules, and those that either did not follow these changes or just chose to ignore them. Even worse, the changes led to even more emphasis on dealing with minor or trivial events without regard to their ultimate potential to become serious incidents. Minor, pinhead-size, second-degree burns counted just as much as a broken bone if the broken bone did not require time away from the job or work restrictions. It also led to tortured interpretations: if an injury was caused by one breath of a hazardous chemical it would be treated as an injury. If two breaths were taken, and the employee exhibited symptoms, it would be recorded as an illness. This became important because *all* illnesses were recordable under the OSHA rules then in place. These rules specified that recordability was determined by the type of treatment provided. You can imagine injured employees' responses when a supervisor or safety representative asked if they had taken one breath or two while exposed to the material that injured them. Savvy employees learned to answer one to try to avoid having a recordable injury and the resulting effect on the company and on their personal record. How hard do you think the supervisor taking the report would push to be sure it was not two breaths when his performance review (translates to promotions, raises, and bonuses) were frequently based on the number of recorded injuries and illnesses? Do not miss the phrasing here. These performance measures were based on number *recorded* not the number that were *recordable*. Learning the letter of the rules of recording became at least as important as learning why the accident occurred in the first place.

Because the pressure to reduce accident rates often remained, this was a conflict scenario for many management groups. Should they count everything and protect themselves from a potential OSHA citation? Or should they focus on recording only those that were obviously recordable to protect their incentive pay or bonus? Or should they keep two sets of records. One for OSHA and one for internal purposes, such as determining how bonuses are paid?

Of course, a result of the conflict is that differences abound between companies and even different sites within a company. One company records all second-degree burns. Another records only those with blisters over the size of a dime. Another records only those burns with blisters the size of a quarter. Yet another records incidents only when they require medical treatment. One employer records all allegations that an injury was work related, while another requires clear verification that an injury was work related. Yet another records only those injuries that are also recorded on workers' compensation logs.

Differences in recording create a problem when companies seek to find benchmarks. When benchmarking on safety, the accident rate is often used to determine who has the best safety performance. These benchmarking efforts seldom consider the wide variation in criteria discussed above.

INCIDENT-BASED SAFETY MEASURES HAVE THEIR GOOD AND BAD APPLE COUNTERS, TOO.

In addition to the problem of inconsistent criteria is the problem of intentional distortion of the measure. Virtually any measure is subject to manipulation. Dr. Deming provided frequent examples of this. One of the simplest was in an apple-sorting process. The measure of performance was the number of good apples sorted. They put the good apples down a chute that had a counter to keep track of the number sorted. There was a control to protect against cheating in that they routinely audited the apple lines to be sure the sorters were not putting bad apples through the good-apple counters to raise their score. On investigation, they asked a particularly open apple sorter if there were ways to beat the system, to which she said, "surely," and reached into the good-apple chute and flicked the counter switch fifteen or twenty times. No apples had gone through, good or bad, but the sorter had increased her score. She further relayed that workers kept an eye on how they were doing, and if they fell behind their quotas, most, if not all, knew this trick to keep up. Incident-based safety measures have their good and bad apple counters, too. All too often the criteria for recordability vary directly with the proximity of the site to their goal. If a site is very close to achieving a goal, they will often go to extraordinary lengths to be sure they can meet it. This is especially true if significant rewards or punishments are placed on achieving or missing the goal.

If employees know they will get a safety bonus if they meet their goal, the minor injuries they would normally have reported go unreported. The burn that occurred at work suddenly happened on the grill at home. As the stakes raise, so does the desire to win, along with the willingness to cheat. Managers begin to overlook these little white lies. They become active in the process to manage the recordability of accidents. This is the reason behind the stories of employees with broken legs who are picked up by coworkers, their boss, or even in a taxi and transported to work. They are assigned to sit with the guard until they can no longer stand the pain, then are taken back home. The result may be no lost-time injury, but it is also a loss of information to help prevention.

Examples of manipulation of incident reporting we have seen include:

- Use of over-the-counter ibuprofen instead of prescription dosages. If the prescription is not used, then there is no requirement to record the incident. Many doctors will cooperate with this approach. This cooperation is rationalized by the assumption that this is what the employee or employer wants, that it provides the same relief for the injured employee, and that it is a less expensive approach. In the end a metaphorical good-apple switch is flipped when a bad apple was found.
- First aid is given for more and more serious injuries that could have or, in some cases, should have had professional medical treatment.
- Preexisting conditions are interpreted liberally to avoid recording an incident or to count a case as one case instead of two. Sometimes the first injury or illness case will have been in a different year so, even if they count it, the first incident is not included in the measures for the time being considered. It can be recorded with no negative effect, because most companies do not factor in such historical changes.
- Employers attempt to classify more events as not being work related. The pendulum swings from taking the employee's word for a case being work related, to one of requiring the employee to prove the case is work related beyond any doubt.
- Employees are offered full pay to work at home when their injuries prohibit working at their normal workplace or traveling to and from their workplace. The requirement is that they go along with the story that the injury did not result in lost work.
- Employee job definitions are used to define the work relationship. In one case we are aware of, an employee fell from a scaffold, breaking both wrists and suffering multiple other injuries. The injuries resulted in a hospital stay, but the case was not included on the injury log, because it was argued the employee was not doing his regular job. It seems this was an infrequent task that was not in his regular job description.
- Medical diagnosis is manipulated. An employee was cut and received fifteen stitches. After consulting with a doctor, it was argued the stitches were cosmetic not therapeutic, and therefore the case was not counted.

These are only a few of the many approaches used to avoiding counting. It makes one feel there may be some truth in this old joke:

> *An employee fell from a rooftop and his scream drew the attention of his supervisor. Lucky for the company he screamed, because it saved their twelve-year, no-lost-day case record. It seems the supervisor was able to fire him for a safety violation before he hit the ground and was injured.*

Manipulation comes against a regulatory backdrop that creates widespread cynicism among safety professionals. In their own interpretations, most in the form of formal interpretive letters, OSHA had required recording these kinds of cases:

- An employee alleged symptoms because another employee had put a voodoo hex on him.
- An employee reinjured an ankle at work but could continue work. This became a lost-time injury when that same employee went to the beach on vacation. The sand and sun caused an irritation under the gel cast and resulted in a rash. Because of this, the employee could not wear his support and therefore could not go to work. Thus, the cause of the lost time was a good time in the sun on the beach. But the local OSHA office determined it to be work related and recordable as a lost-day case.
- Any blister caused by a burn is recordable even if it is the size of a pin head.
- It has been argued that a cold that an employee alleges was caught at work would be recordable under the OSHA requirements.
- A mild sunburn, since it is a symptomatic response that is not the result of an instantaneous exposure, could be recordable.
- One breath of a harmful substance constitutes an injury, while two or more breaths constitute an illness.

OSHA had more than 200 such interpretations under the old rules, and the new rules will undoubtedly generate a new set of interpretations, which will further erode the utility of this common measure of safety. Is it any wonder employers first bend the rules by not recording extreme examples like these and then gradually accept cases they would not have thought of leaving off the records before?

Understanding these issues, how can two companies compare their statistics and believe the numbers provide an indication of the relative quality of their programs? The company with the higher reported record may actually have a better safety culture than the company that has a lower accident rate, because the latter company is hiding accidents when they occur. The higher rate may be an indicator of a better process by virtue of the fact that they are identifying and recording the accidents that happen. Or it could simply mean they are having more accidents.

The OSHA record-keeping criteria have been discussed in detail as an example of the issues with all of the injury record keeping we reviewed. The old ANSI system had just as many different ways of using the same criteria as does the OSHA system. The international schemes we have reviewed have similar problems, with the added complexity of vastly different cultural attitudes about accident and prevention measures.

The problem noted here is not so much that employers are scofflaws and do everything they can to avoid the record-keeping rules out of contempt or malice. Frequently managers might not even know when marginal reporting was taking place. In almost every case we are aware of, the companies went to great efforts to prevent unlawful practices and routinely disciplined those who knowingly committed these acts, including terminations in some cases. The problem is also not so much the fault of OSHA being an agency run amuck. Most of the record-keeping interpretations mentioned above were made with good intentions and in response to real incidents. OSHA was only responding to the need to clarify and manage a system that inherently cannot be managed to the fine standard that was being asked of it.

A serious problem is that the events being measured are so complex that systems to define and count them must be elaborate and cumbersome. In addition, there are often strong incentives attached to the numbers achieved. This complexity provides more and more opportunity for errors or manipulation, both intentional and unintended. Although the regulations may stand as an attempt to create a useful operational definition of a recordable accident, we are convinced that, in practice, this has not been entirely successful. Moreover, the attempts at creating simple definitions have contributed to a focus on trivial events and a failure to deal with events that portend future catastrophe but do not meet the criteria for recordability. Finally, the incentives for lowering the accident rate often provide a strong inducement for manipulating the numbers.

Variation in the Statistic Itself

Even if we accept that the recording process in a particular population could be reliable, we are still left with the fact that the variability of incident rates is quite high for small groups or for short periods of observation. As we discussed in Chapter 2, the appropriate statistic for computing the standard deviation of an accident rate is the *U* statistic, based on the Poisson distribution. The standard deviation,

$$U = \sqrt{\frac{R}{N}}$$

where *R* is the recordable rate and *N* is the number of 200,000 work-hour exposure units. The 200,000 hours is the expected annual work of 100 workers.

When we compute control limits of a process that has an average OSHA recordable rate of 3, the range of the control limits increases greatly as the population decreases. For a plant with 25 workers, the standard deviation,

$$U = \sqrt{\frac{3}{0.4}}, \text{ or } 2.74$$

The upper control limit is three standard deviations from the mean, or 3 + (3 x 2.74), or 11.22. This is based on one year of data. If the observation period shrinks to a month, then the units of exposure go from .4 to .033, and the upper control limit rises to 31.60. This looks high, but how many incidents can happen in a month for the measure to remain within control limits? The answer is that one accident in a month will yield a rate of 30.3 for that month, just inside the control limits. At this site, if we had two recordables in the month, we would exceed control limits. The point here is that comparing either the recordable rates of small sites or rates of larger sites computed over short periods is not very useful. If you measure two plants with 40 employees for over a year and one has a rate of 8 and the other of 0, there is no statistical basis for concluding they are different. This difference is within the normal random variation of measure being used. The incident rate of 8 took only 3 recordable events to reach.

Figure 4.1 shows how the control limits change as the plant population changes. The mean recordable rate is 3. The control limits are for one year of data. Monthly control limits would be much wider.

The meaning of this is important. Safety managers would generally conclude that a recordable incident rate of 1 was good performance, whereas an incident rate of 10 was not very good, and a rate of 30 was terrible. With a plant of 25 persons, an underlying rate of 3, and one year of observation, we

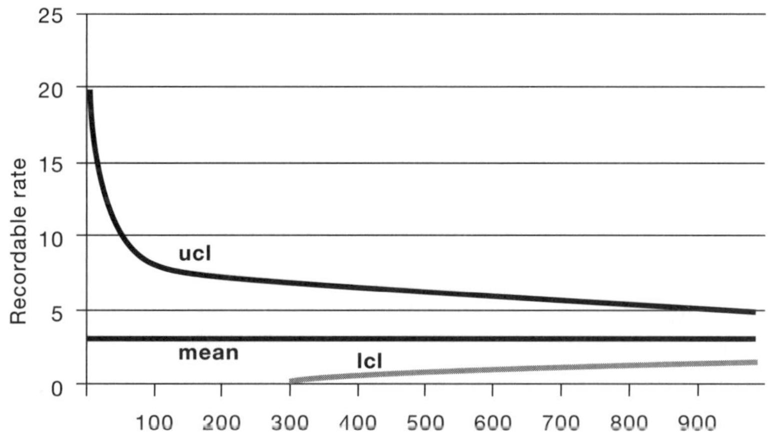

Figure 4.1 Control-limit changes linked with population changes.

might observe a rate of 1, or we might observe a rate of 10. In fact, we might see a monthly rate as high as 30! Any of these rates would be within the control limits and therefore would not signify that the process had changed. Remember that these observations are based on the same underlying safety process. If we are aware of the statistics, we will need several years of data to draw a definitive conclusion about the safety performance of this plant.

VALIDITY

We know of only a very few companies or safety professionals who have questioned the validity of incident rates as a measure of safety and even fewer who have done anything about it. However, it is important to recognize that all measures have limitations on validity. Moreover, a measure like incident rates, which has low or, at best, moderate reliability must, of necessity, have limited validity. It is important to discuss the limitations on validity.

Content-Related Validity of Incident-Based Measures

Accident rates have obvious face validity as a measure of the safety process. After all, most recordable accidents are instances of something we are attempting to prevent. At the same time, a cut finger, on its face, is very different from the release of a large cloud of toxic gas. To the extent that we are concerned more about the toxic cloud than about the cut, the cut finger may have little content-related validity. If the release does not result in an injury, in many current measurement schemes, the cut finger will receive higher importance value and demand more resources be directed toward its prevention. This scenario is one that first pushed us toward our research for this book. It made no sense that the most widely used measure was leading us to make what appeared to be misdirected decisions. It might be argued, as Heinrich essentially did, that the release and the cut finger are both instances of exactly the same thing, and therefore cut fingers (incidents) are a measure that is valid in all respects. One could make an argument that all incidents are evidence of management-system failure and should be investigated. However, in an environment in which resources are not unlimited, incidents that are related to potentially serious outcomes should be investigated first, whether they happen to be recordable or not.

Realize that the relationship between minor incidents and serious incidents is an empirical question, and no amount of argument or opinion can settle it in the absence of data. Figure 4.2 depicts Heinrich's (1959) assertion that minor incidents, moderate accidents, and major accidents are causally related. However, to show that the ratio of minor injuries to major injuries is 29:1 or something similar does not in any way demonstrate that they derive from the same process, or that reducing minor injuries will lead to fewer major injuries.

For example, consider near misses, OSHA recordables, and lost-work cases. A paper cut would be a cut and an injury but not worthy of a lot of attention. Cutting off a fingertip with a paper cutter is important and should be studied. However, a minor cut in a powered paper-sheer machine that could have easily resulted in the loss of an entire hand should receive considerably more attention and resources to investigate and prevent.

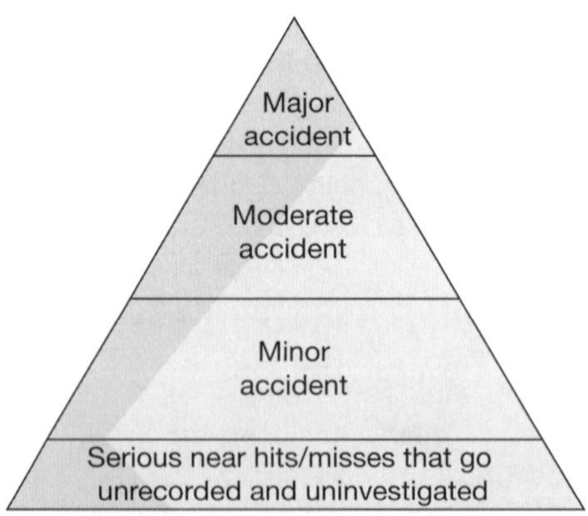

Figure 4.2 Revised theoretical relationship of minor and moderate accidents that have the potential to yield major accidents.

Treating all events the same and assuming that by preventing minor incidents you proportionally reduce the likelihood of more serious events causes organizations to attack the easiest causes first. This frequently draws attention and resources away from finding and correcting the system hazards that could produce very serious outcome such as the chemical release that is overlooked because the company was too busy investigating steam tracer burns.

Criterion-Related Validity of Incident-Based Measures

Accident rates are frequently themselves used as a criterion. However, safety is not necessarily the same as an absence of accidents. Dictionary.com defines safety as *freedom from danger, risk, injury*. Conditions are easily conceived in which there is no history of injury but great risk of future injury. Of course this appears to be the case with shuttle flights up to the time of the Challenger disaster. Even though there was no history of injury, even the lowest estimates of danger would have indicated that shuttle flights were quite risky.

Perrow (1984), in his theory of normal accidents, argued that what he called normal accidents, disruption of the flow of complex processes, were related to potential catastrophic events. Most of these disruptions in flow would not be classified as recordable accidents. In fact, plants that present the most significant risk to workers and the public are typically high-capital, non–labor-intensive plants, likely to have very low rates of recordable accidents.

In a test of Perrow's theory, Wolf and Berniker (1999) studied data on hazardous releases at 36 refineries from 1992 to 1997. Their assumption was that reportable releases are instances of normal accidents. They also provided data on the total case rate and catastrophic incidents. There were three fatal accidents in the period. They argued from the data that the incident rate was a poor measure for predicting catastrophic loss. Better predictors were: (1) The complexity of the system. More complex, tightly coupled (continuous-processing) systems have the highest danger. (2) The number of toxic releases. The data showed that these releases were much better indicators of system problems than were recordable accidents. Although there was a small positive correlation between case rate and releases ($r = .23$, $n = 22$), it was far from statistical significance ($p = 0.30$, two-tailed). The relationship between case rate and catastrophic accidents is not statistically significant, whereas the relationship between releases and catastrophic accidents is very significant ($p < .01$).

Incident measures certainly have some criterion-based validity. Because they represent injury and cost, they are themselves a criterion. They are not, however, the only important criterion, nor are they a substitute for measures of catastrophic loss or for measures that more accurately forecast the probability of such loss. To gain perspective on this, consider how much time in your safety program is spent dealing with the serious hazards you face, compared with the time spent dealing with minor or trivial hazards. Are you satisfied that the serious hazards are receiving enough attention? Do the minor events command the lion's share of your time? Are the rules of accident recordability setting priorities for your safety process?

Construct-Based Validity

To consider construct-based validity, we have to address the question of what it is that accidents are measuring. This moves us to the consideration of accident causation. The logic is that accidents are providing us with information about a set of underlying processes that we label causes.

Heinrich suggested that in 85–93 percent of cases, the primary cause is a human cause, so that incident rates may provide us with a measure of unsafe behavior. In fact, it is very likely that 100 percent of all accidents involving anything manmade or involving people in any way involve a human element[1]. In some way a human decision, action, or inaction was a contributing cause of the accident. It is just as true to say that, as we define the component parts of the accident-producing process, each of the parts makes at least some contribution to the final event. Each part is a theoretical contributor to nearly 100 percent of the cases, and not just the 85–93 percent suggested. If the accident-producing process is defined as having the three major components—human, material and organizational—it becomes easy to show that each must be considered a contributor in every accident.

The human contribution must also be considered on a much broader basis than simply human error. Physical limitations, natural human tendencies, cognitive limitations, ergonomics, knowledge, skills, and other characteristics of the people part of the accident-producing process must be considered. This is especially true if we are to identify fundamental and permanent improvements to the process so it produces fewer accidents in the future.

Because accidents are the product of the system, no part of a system can exist without influencing every other part of that system in some way. When people are part of the process or system that produced the accident, they must have, in some way, contributed to the accident's happening as it did when it did. The contribution can be conscious or unconscious, intended or unintended, direct or indirect. The contribution is not necessarily fault. It is simply part of the system of causation.

To take this position requires expanding our thinking to accept the idea that not only are people an element of almost all accidents, so must the other components of the process or system be a part of every accident. The machines, methods, materials, inputs, outputs, work environment, organizations, management, agencies, and every other element of the system related to the accident are parts of the system producing the accident. These elements may be grouped into the three major categories.

Human. The human element consists of actions or failures to act based on knowledge, beliefs, attitudes, physical abilities or capabilities, motivation, and so on. Usually the actions of operators are listed as a principal human cause. After all, they are there when the accident occurs, and their actions are closely connected in time with the accident. However, it is more appropriate and more productive to look at the entire community of humans who influence the event, including some who are no longer employees and some who might never have been employees. We would categorize the roles of these people according to Figure 4.3.

The internal lines are dotted because the distinctions are not necessarily sharp. Individuals can have both managerial and technical responsibility, for example. The staff/technical category includes the designers of the system, who may have departed from the company a long time ago or who may have never been employees. Other similar circles can be added for peripheral or indirect human-cause groups, such as agencies, customers, suppliers, family, and the like.

Consider the disaster at the Hyatt Regency Hotel, in Kansas City, on July 17, 1981. Two suspended walkways crowded with people collapsed in the lobby of the hotel, killing 114 people and injuring nearly 200. The circumstances are well described in Henry Petroski's important book, *To Engineer Is Human* (1982). The extensive investigations that followed

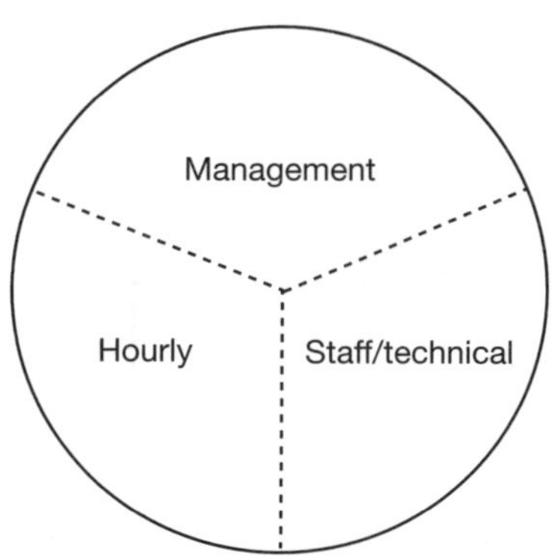

Figure 4.3 The roles of people contributing to an accident.

revealed that the original design of the walkways, though structurally sound, was highly unusual. The concept was that the two walkways, one below the other, would be suspended from a set of rods hung from the ceiling. Each rod was to pass through the upper walkway and secure both walkways. This meant that the stress on the ceiling fastening point would be equal to the weight of the two walkways and the people who stood on them. The joint at each walkway would have to carry the load of only that walkway. However, it was rather difficult to imagine how the structure would be assembled. Proper assembly would require passing a long rod, perhaps threaded for much of its distance, through the upper walkway, starting a nut onto it, and running it all the way up to secure the upper walkway. The lower walkway would then be secured to the rod by a second nut. This is depicted on the right portion of Figure 4.4.

Apparently, the architect did not specify how the assembly was to take place. The contractors on the site decided to carry out a much simpler process. They suspended the upper walkway from rods coming down from the ceiling. The lower walkway was suspended from rods secured on the upper walkway. This is depicted on the left side of the diagram. Unfortunately, the contractors failed to grasp the physics of what they had done. What might appear to be a design change of minor consequence actually had the effect of suspending both walkways from the joint that secured the upper walkway to the rod coming down from the ceiling. This is circled in the diagram. Surely the architect who designed the structure would have understood this, but he or she was apparently not there to supervise the construction.

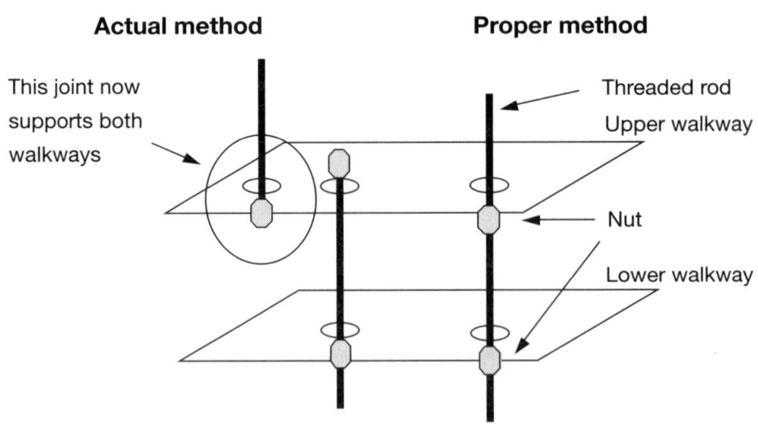

Figure 4.4 Design of walkways in Hyatt Regency Hotel disaster.

During the construction process, the workers noticed that the walkways seemed to be very unstable when they pushed loaded wheelbarrows across them. If they complained, however, nothing was done.

When the walkways were finally loaded with people attending a celebration and swaying to music, the joint securing the upper walkway failed. The two walkways collapsed onto the crowd below. The casualties were so extensive that virtually everyone in Kansas City knew one of the killed or injured.

What was the human cause? Surely we can identify the architect, who offered a design that would be difficult to build and who failed to ensure that the design was carried out faithfully. Of course we can identify the contractors, who made design changes without appreciating the engineering consequences of their actions. We can also identify the construction workers and their foreman, who failed to make an issue of the structure's instability. Perhaps there were others as well. Interestingly, the proximal cause of the collapse was dancing on the walkways, but we would not assert that the people dancing were in any way culpable. Not one of the persons who had a critical role in the collapse was on the scene.

Material. The material element consists of those hardware or physical aspects of incidents, such as wrong construction material, oil on the floor, failed relief valve, plugged line, and so on. This includes the physical work environment, working conditions, equipment, supplies, and materials. Elements include lighting, temperature, materials used, housekeeping, ergonomic aspects of the task, guarding, equipment, and other physical elements of the working environment. Based on the work of Wolf and Berniker (1999), we could also include the complexity of the physical environment. Again, these are in large part the results of human activity. But like culture, they have a life that is somewhat independent of the persons currently involved in the system and are therefore worthy of identification as a causative factor. Because 100 percent of all accidents involve a human element, an organizational or a cultural element, and a material element, systems theory tells us that one element cannot be changed without affecting the others. Nor can one element cause an accident independently of the rest of the system.

In the Hyatt tragedy, one material factor was a design that was difficult or perhaps even impossible to manufacture.

Organizational factors. These relate to the structure, practices, procedures, policies, and so on, in place in the organization, as well as its history. Policies and practices include hiring policies, compensation policies, promotion policies, reporting structures, training practices, work standards, regulations, and enforcement practices. They may be stated or actual policies and practices, that is, what is actually done as opposed to what is written or said. The obvious organizational factor in the Hyatt tragedy was the failure of communication between the designer and the builder.

Culture of the organization. The human, material, and organizational factors come together to form the cultural system of the organization. The cultural system represents the attitudes, beliefs, and expectations of the people in the organization. These are the product of human experience over time with the organizational factors and the material. All events are interpreted through the lens of this culture. In spite of changes in emphasis, rules, procedures, incentives, and personnel, the culture is resistant to change. For example, in organizations that have a history of short-lived initiatives that are soon replaced by new initiatives, employees come to expect the flavor of the month. Thus, any new initiative is greeted with cynicism and resistance that will be passive at best and active at worst. Change is stalled, and the expectation of short-lived initiatives becomes a self-fulfilling prophecy. Changing culture requires changing the experience of the people over time. An organization in which senior management has paid little real attention to safety cannot expect to see an immediate change when some attention is given. However, if that attention and involvement of senior management is maintained over a period of time, on the order of years not months, the culture will change.

Although the culture is hard to change, it is also a source of great leverage. Culture can often overcome momentary lapses by individuals and occasional problems in the material and organizational systems. At the simplest level, this could involve one employee telling another to put on safety glasses or to slow down when observed driving too fast on the plant site. At a deeper level, it would involve shutting down a production process when it appears to be operating improperly. This latter act is very difficult for an hourly employee to do without a supportive safety culture. Thus in a strong, positive culture there will usually be quick identification and action when problems arise. If violations of safety regulations, hazardous physical conditions, or unsafe procedures are never part of the way of doing business, they will not be tolerated in the culture when they are discovered.

In the Hyatt tragedy, we do not have enough information to speculate about the culture. One typically learns about a culture by analyzing the root causes of numerous incidents or conducting interviews and surveys.

Our system of accident causation is diagrammed in Figure 4.5, with the three areas combining to create an organizational culture.

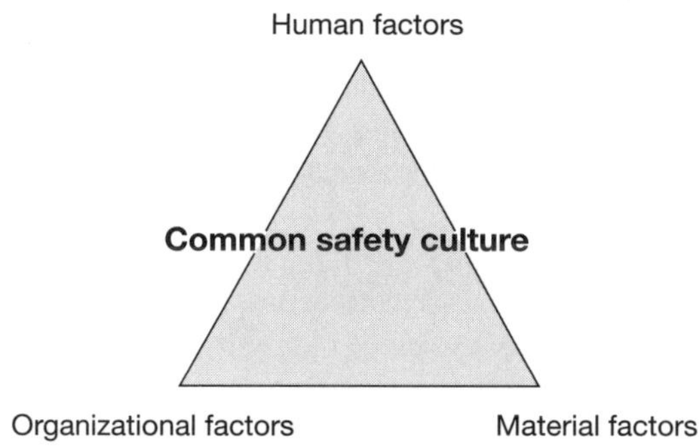

Figure 4.5 Factors in accident causation.

To see these sources of causation at work, consider the following incidents:

Case 1: At the end of a production run, an employee transferred a batch of the product to a storage tank. The employee did not follow procedures. He failed to take a gage reading on the tank before the transfer. The receiving tank did not have enough volume left to hold the transfer. When the tank became full, the internal pressure built, and the bottom seam failed. The bottom of the tank separated from the sides, and the contents spilled onto the ground.

A traditional view of this accident would be that the employee had not followed established procedures and had caused the accident. It is an open-and-shut case of human error, one of the 93 percent we knew was out there. Because they believed that most accidents are the result of human error, the investigators looked for the human cause first, and when they found it, they concluded their investigation.

Later we reconsidered this same accident as a system of human, organizational, and physical elements. Investigators began to look for the other elements and how they influenced the accident. What they found was that several questions had been left unanswered. A critical one was, "Why did the tank rupture instead of overflowing through the relief valve?" On further review, they found the relief valve had been taken off the regular test and inspection list. When tested after the incident, the relief valve did not open due to corrosion of its internal parts. This maintenance failure then points to a management problem. The system of supervision and control had failed to maintain the integrity of an important safety system. Thus we discovered an organizational factor. The staff/technical group also failed in that the original design and engineering systems had not recognized the possibility of corrosion in the type of service experienced by this valve. The physical systems played a part in that the working parts of the relief valve had corroded and had not allowed the valve to function properly. There was also a physical reason the valve went unchecked. The valve was located where it was very difficult to reach. Over time, someone on the maintenance crew had apparently removed the valve from the inspection list. This might have been done to avoid the unpleasant work of testing it every year.

This accident was a result of interactions among physical conditions, human actions, and organizational systems that failed to ensure regular inspection and maintenance, and perhaps a culture that did not instill the importance of such equipment to the group that maintained it.

To correct the problem, in order to prevent any similar accident from occurring, would require addressing all three elements. A simple reprimand of the operator for being careless and not following procedures, no matter how severe, would be entirely ineffective. More likely it would be counterproductive. It would increase alienation between management and hourly workers, create fear, and slow the flow of communication that is essential to safe operation. Perhaps the most leverage would come from the organizational and cultural issues of maintenance. A very important question would be to find out whether similar maintenance failures were found in investigations of other incidents. If there were a pattern, it would be a very important place to take action.

Case 2: A gasket failed in a line transferring a highly flammable liquid. The liquid was released, and an explosion and fire ensued. The gasket met the design specifications for the service and had been installed properly. Upon installation all testing had been completed according to procedures.

On initial investigation, this was a physical cause. The gasket failed. Upon considering this as a system, it was found that the gasket was not proper for this service. The pressures on the system were high enough to cause the gasket to deform slightly every time a transfer was made. The original testing did not find the problem, because the problem was only manifested after repeated pressurization. The three pressure tests during the commissioning of the line were not enough to cause the failure. The cause found by a systems review was that the specifications were not appropriate for this service.

One human element had been that the engineer (staff/technical) writing the specifications had not considered the creep factor of pure

Teflon. He had only considered the corrosion and temperature characteristics of the gasket. He had not known, or had not thought to consider, the effect on the gasket of repeated pressure fluctuations.

An organizational element was that the specifications could be written without adequate consideration for in-service conditions. This specification had been completed and included in official specification manuals with this serious error. The error was made out of ignorance of all of the parameters that needed to be considered for this particular service, and it was subsequently not caught by what was considered to be a stringent specification review process.

In this case the human and organizational elements were not only spatially distant from the accident but also temporally distant. The specifications had been written at a different site some six years prior to the accident. If other incident investigations found similar specification problems, there would be a legitimate concern that the system that created these specifications was not functioning properly. Then it would probably be a good idea to review all critical specifications, although this would be a huge task.

Case 3: The regular transfer truck driver was out one day when a trailer load of a product needed to be moved. Knowing that drivers from another part of the plant were available, a supervisor asked one of them to move the trailer. A driver came across the plant with his transfer tractor, hooked onto the trailer, and pulled away. No one had disconnected the trailer from the loading line, and when the driver moved the truck, the line broke. The product was released.

On initial investigation, this might have been considered a management failure. The supervisor had asked a driver to do something he had not been used to doing. The management process had not assured the availability of experienced drivers for the unit that needed them that evening. Nor had it required confirmation of appropriate training before undertaking new tasks.

On viewing it as a system, it was found that although the driver was from a different area, he had been trained to check each trailer for wheel blocks, disconnect fill lines, and check safety-valve positions before ever moving it. This driver was in a hurry, thought the trailer had been properly prepared for transfer, and did not bother checking the connections. This was an important human element or cause of the accident. Taking this to the next level, we would ask, "Why was the driver in a hurry?" and "Was the culture of the site one that would routinely accept his taking these types of risk?"

> On later testing it was found that the excess-flow valve on the trailer had also failed. When the line broke, the valve should have closed and limited the release of the tank contents. It did not, and considerably more product was released than would have been had the valve functioned properly. This was a physical element or cause. Furthermore, it was later found that some modifications made on the excess-flow valve, made some years before, had caused this physical failure.

At this point it is important to note that the effect of the different factors is not always negative. For example, quick response by an employee in an emergency may prevent or limit the extent of the damage from an accident as well as contribute to it. Safety factors built into many physical systems often prevent accidents even when design specifications are exceeded. Redundant safety systems and training required by an organization (internal or external) for operations and operators can also result in avoiding accidents or limiting their effects.

In order to use data from incidents for system improvement, it is necessary to understand their systemic causes. Repairing the local causes of a specific incident provides no leverage. If a single worker trips on a particular stairway, then putting nonskid pads on the steps will have a negligible impact on future incidents. If we discover that he was in a rush to get to another area of the plant, and if we discover a number of similar incidents caused by rushing, and if we discover that there are some fundamental work-scheduling problems that we can address, then we have gained considerable leverage.

If a site is spending all of its resources finding, investigating, and defining actions to correct minor issues revealed on specific incidents, then there are little or no resources for acting on removing major system hazards. Do we pave over parking lots because someone stumbled in the loose gravel, when important safety interlocks in the workplace go without repair due to a lack of funds? Do we spend time and resources replacing first aid kits because they were not physician approved, while improper electrical installations go unchanged in potentially flammable or explosive atmospheres? Has it happened? Is it happening now? Will it happen in the future? You bet it has, is, and will.

Probably the most important responsibility of management is to set priorities on the use of resources to affect the most meaningful issues the company faces. A problem that the success of the safety profession has generated, however, is that managers believe they get measured on what is important to the company. Many companies evaluate staff based on injury rates and number of corrective actions initiated. This provides an incentive for nonreporting and for taking action on the most trivial matters.

We are having great achievements climbing the safety ladder to success. The trouble is that all too often the ladder is up against the wrong wall. When we get to the top, we are worse off than we would have been without the success. Now we have to retrace our steps and unlearn all of those things that had made us a success or have the courage to jump from one rooftop to another. A real example of the effect of this theory of reporting and acting on every incident or potentially hazardous condition is relayed in the following story.

Case 4: A large site subscribed to the concept that the more hazardous conditions that were reported, the more that could be investigated, and thus the more accidents that could be prevented. The site installed an elaborate system to collect reports, complete with graphs and management measures and an attractive incentive program to get more of the reports submitted. The success of the program was phenomenal. In no time at all, they were receiving hundreds of incident reports and reports on hazardous conditions. They were also doing their level best to respond to each as a demonstration of their commitment to a safer workplace.

All seemed to be going well until all of the obvious easy-to-fix problems were identified and corrected. To keep up with what had become a quota of expected reports, the site literally had reports like "toilet paper too rough" submitted as a hazardous condition. This report was submitted anonymously and obviously to make a point. Everyone began to realize they were reporting for the sake of reporting and to meet the reporting levels required to be eligible for the incentive program, not for the sake of identifying and correcting hazards.

During the initial stages of this program, incident lost-day case rates stayed high, even rose above previous years, while recordable cases fell. In retrospect, what was most likely occurring was that cases that could be hidden were being hidden, and those that were just too serious to hide were being recorded. By focusing on reporting, measuring, and acting on all events, the site had lost its ability to distinguish between the important and the trivial. Managers and nonmanagers alike were intent on measures for the incentive program and lost sight of what the incentive plan was intended to effect.

Of course, the problem here is that all actions taken were local and not leveraged. System causes were not identified. Instead, we have a flurry of meaningless activity and wasted effort. It is a leap of faith for a traditionally trained safety or environmental professional to say we will stop spending time trying to alter conditions that have no potential for producing serious events and focus our effort on those system causes that do have that potential.

WE HAVE TO BE ABLE TO PRIORITIZE WHERE OUR RESOURCES AND EFFORTS ARE APPLIED.

We have to be able to set priorities about where we apply our resources and efforts. No matter what levels of resources are available, it is still important to decide which is the first best step to take. This is not reducing the importance of addressing all hazards. It is simply taking a rational approach to where one begins and how one plans to progress through the list.

In this case, when we began to look into the organizational culture, we found that the most significant barrier to change and improvement was fear. Further analysis indicated that a key demonstration of management's desire to improve the system would be their willingness to revise how they conducted accident investigations. A retired union leader told us they had coached their members to respond only to questions asked and never to volunteer any information. They did this to protect themselves and their union brothers and sisters. He relayed that the company had missed many of the real causes of accidents simply because they had not asked the right questions.

On reflection, the site management began to understand that they had always looked for the human factor (a human-error scapegoat) first. To compound the problem, the human error found was almost always the nonmanagement or nonsupervisory one. They were not identifying the supervisory and management impact on the processes producing the accidents, so they were not taking action on them. The union people were likely justified in holding back, because they were always found to be the problem.

As a long-time safety professional, I am acutely aware of the difficulty of accepting that some accidents are just not worth the time it takes to investigate them, and that some accidents should not be given corrective actions. Some events just have to be accepted as part of the process of living and doing business. Some risk must be taken in order to deal with much larger risks that might otherwise be ignored. I know this is safety heresy. I know many who have long extolled the notion of zero accidents as an ultimate goal will recoil from the idea that we are better off accepting some minor accidents and focusing on preventing the more serious accidents. I know some will want to have my certified safety professional (CSP) revoked. That was my response when I first heard this suggested by a keynote speaker[2] at an American Society of Safety Engineers (ASSE) professional development conference in Boston several years ago. Over time I have come to agree with this seemingly outlandish idea.

In attempting to summarize the issue of construct validity for incident-based measures, we can say first that we are measuring the gage itself, or the incident-recording process of a particular unit. Second, we are measuring the output of a very complex accident-producing process that includes the people, physical environment, and organizational system. Some incidents may reveal problems with these processes that could produce catastrophic outcomes. Other incidents may offer no insight at all into such problems. To the extent that incidents are what we want to prevent, we need not worry about this limited validity. However, if our primary objective is to prevent major or catastrophic loss, then we need to be very conscious of finding systemic causes and directing our actions first at the most serious hazards.

PROPER USE OF ACCIDENT STATISTICS

Incident rates should be plotted on a control chart in order to see if the process is changing over time. The first step is to establish a process mean, which would be the mean recordable rate over the past 12 months or so. The next step is to calculate control limits. The upper control limit is equal to the mean plus 3 standard deviation units as calculated by the *U* statistic. The lower control limit is the mean minus 3 standard deviation units. Of course it cannot go below zero. The mean and control limits are laid out on a chart, and each month the recordable rate is plotted. The following events indicate that the process has changed (Nelson 1986):

- A single point above or below the control limits
- Two out of three consecutive points in the upper third of the chart (more than two times *U* from the mean) or two of three consecutive points in the lower third
- Seven consecutive points above the mean
- Seven consecutive points below the mean
- Six points in either ascending or descending order

Should any of these criteria be met, it is appropriate to try to find out what happened. If you have a run of points above or below the mean or several excursions outside of control limits, it is appropriate to compute a new mean and control limits, going back to the time that the run started or the first out-of-control point was observed.

As long as the process stays within limits, it is not useful to question what went wrong or what went right in a particular month. You are simply observing random variation. Trying to track down the cause would be like trying to find out why a roulette wheel came up on 17 on a particular spin or hit red four times in a row. On the other hand, if the wheel hit red eight times in a row, investigation might be worthwhile. That is the logic of the control chart. We should look for the cause of a variation only when the process varies in a way that is very unlikely to have been due to chance.

It is critical that normal fluctuations of the measure, within the control limits, not be treated as signs that the process has changed. Acting on these random fluctuations as though they represent process changes is called *tampering*. Tampering not only wastes resources but usually makes the performance of the system worse.

Examples of the Use of Control Charts for Recordable Accidents

Consider the case of an organization that has an average of 6,000 employees spread across three organizational groups. Group 1 has averaged 4,000 employees; group 2, 1,800 employees; and group 3, 200 employees.

Figure 4.6 represents a *U* chart for the entire population of 6,000. What is plotted is the recordable rate, month by month. The control limits represent the mean rate plus or minus three times the standard deviation *U*. At the beginning, there is a run of 29 months, during which the monthly

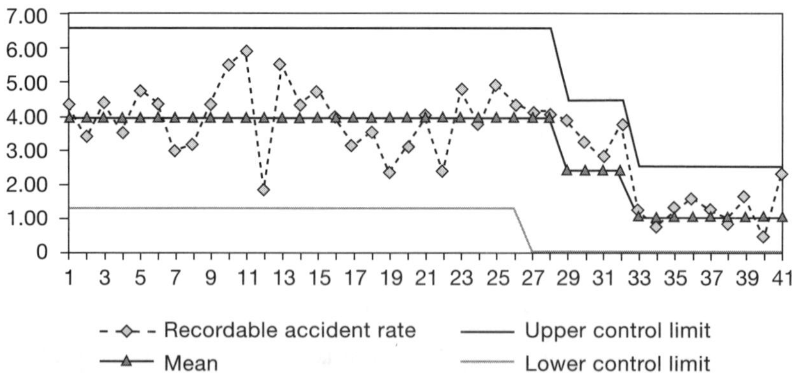

Figure 4.6 Recordable accidents for a combined employee population.

recordable rate varies randomly around the mean of the chart. At month 29, the rate of 3.82 is slightly below the mean of 3.90. This point by itself would signify nothing. However, the next six points also fall below the mean. In control charting, a run of seven consecutive points above or below the mean signifies a shift in the process. Therefore, the mean is recalculated retroactively to the first point in the run. Since the *U* varies with the mean, the control limits are narrowed as well.

Because the data come from history, it becomes even more difficult to find why the process shifted. As the control chart is maintained, it is important to document the potential causes of process shifts that are observed. However, even when good records are kept, it is often not possible to be certain about the cause of an observed process shift. In the organization depicted here, at month 10 on our *U* chart, a major effort to improve system weaknesses began. This effort took the form of starting up several process improvement teams (PI teams). At month 11, after a multiple-illness accident, one group announced an incentive program that would give everyone at their largest site a bonus of $500 if they did not have a recordable accident over the next 6 months. At month 17, all of the supervisors attended a special safety training program. At month 18, an effort to define individual roles and responsibilities (management and nonmanagement alike) was completed and redefinitions communicated and carried out over months 20 through 30. At month 20, one of the old more labor intensive production lines was closed. At month 21, a major accident occurred that got everyone's attention and renewed respect for the risk to which they were exposed. At month 22, a stern statement, out of management's commitment to avoid similar accidents in the future, was sent individually to all employees. It contained the message, "Working safely is a condition of employment at our company. All employees having accidents and their supervisors will report the reasons for the accident and corrective actions to be taken to the executive management of the company!"

If one looked back and considered which of the noted events was the cause of the change in the process, it would be extremely difficult to decipher which had positively influenced the process. The analyst should also be wary of several potential problems. One concern is that the shift is only temporary. If the effect of the heightened awareness of the accident that occurred in month 21 wears off over time, the process could return to its previous level. Another concern is that actions like the "work safe or bear the consequences" letter from management might have caused changes in the reporting process not the safety process. Employees and supervisors, concerned about their jobs and careers, might be more reluctant to report mistakes that take the form of accidents they can hide. This could result in fewer reports and then an apparent process change. This brings up a basic rule of control charting. The first thing to determine when you see a process shift is to make sure your gage has not shifted. In this case, we would want to make sure that the criteria for recording accidents had not changed.

The analyst should also consider possible delayed effects of actions taken months before the shift occurred. The PI team actions, supervisor training, and role-definition actions would all have taken time to show results. The decline in the rate might be the result of those actions showing up coincidentally with the other, more apparently direct, actions. The PI team actions, training, and role definitions would more likely have been long-term drivers of safety-process changes rather than an incentive plan or temporary increase in management emphasis. However, in traditional approaches to analyzing safety data, the proximity of the management letter to the shift would be an easy and self-gratifying explanation of why the shift occurred. This explanation would then promote the same type of action the next time the numbers went up. Such actions as the letter from management cause many cyclical results experienced when management takes a narrow, short-term view of a complex process. Most likely the management letter represents tampering and would slow the improvement process. This does not mean management should not express their commitment to the safety-improvement process. However, it is not useful to instruct employees to work safely. It is clearly a violation of Dr. Deming's point 10: *Eliminate slogans, exhortations, and targets asking for zero defects or new levels of productivity.* Such communications are insulting to workers. The appropriate vehicles for management's demonstration of support are allocating resources to safety and personal involvement in the improvement process. Telling others to make improvements is not an effective form of personal involvement.

Considering all of the issues in this example, it should be obvious that understanding the causes of a shift in the incident rate is very difficult. In fact, the number of events that may have influenced the process is likely to be many more than the number discussed here. The action of many of these factors is indirect. In our example, we may have been experiencing a depressed economy. People in the process might have been afraid of reporting accidents, because they were concerned for their jobs. The economy might also have been good, and the company was able to make some long-needed repairs to the facilities that affected the employees' attitudes and actions. Any or all of these could have affected the safety record.

Referring again to our chart, we see another process shift beginning at month 33, with seven consecutive points below the previous mean. Again, the objective would be to have kept good records of actions taken and analyze those and any other potential causes for the mean to have taken another step down. Whether this is a continuing effect of the initiatives described, or a result of something else, we cannot say. However, it does appear that the company has made some real safety-process changes, and there is good cause to believe the actions taken were having the desired effect.

Going back to our analogy of incident measures being like a rearview mirror, in this case, we can say we looked back and saw that whatever we were doing was yielding improved results.

The groups that make up the chart should also be considered separately. For group 1 we find Figure 4.7. The review of this chart reveals some interesting points. First and most obvious is that at month 11 there was an outside-of-control-limits point. A point fell beyond the upper control limit of the process, or more than 3 standard deviations from the mean. Whenever this occurs, it is very important to try to determine the cause. In Deming's terminology, this is a *special cause*; in Shewhart's, this is an *assignable cause*. According to Shewhart, the probability of a point's being this far from the mean due to chance is so low that it makes economic sense to attempt to find the cause. A point that is only 2 standard deviations from the mean might represent a special cause, or it might represent random variation. However, it does not make economic sense to investigate it, because the probability that it is due to chance variation is too high.

In this case, we believe that the out-of-control excursion was caused by a batch of raw material that met written specifications but was not suitable for the then-current operating parameters, which had been changed in recent months. When the new raw material hit the reactor, the process overheated and blew a rupture disk inside a process building. Several employees were exposed to noxious vapors and were treated for their symptoms.

The supplier was a new, low-cost vendor who met all of the purchasing department's specifications. Purchasing had used this supplier without

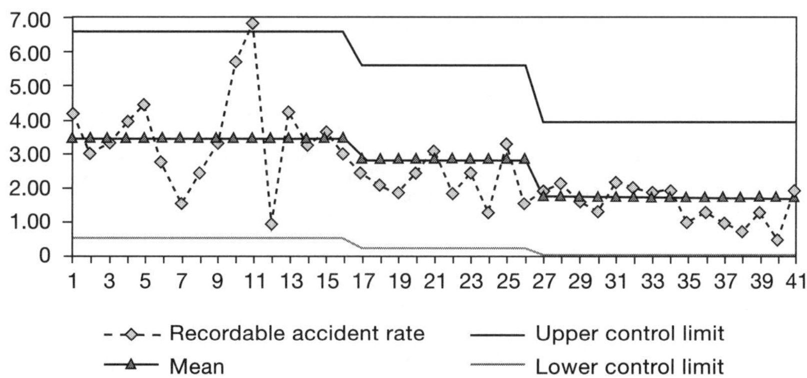

Figure 4.7 Accident rate for group 1.

notifying the production staff. The losses that resulted from trying to save a few pennies per kilo included lost production time, repair cost, loss of two customers who could not be supplied in the newly committed just-in-time fashion, and, most important, nine employees who were exposed to vapors that resulted in illness symptoms. These illnesses were the special cause of nine additional recordable cases. This was a unique event, isolated in time, not a shift in the process. Specific actions were taken, and controls were put in place to positively prevent recurrence.

It is important to note that the statistical basis of the *U* chart is the Poisson distribution. One of the underlying assumptions is that it describes incidents, not the number of persons injured. An incident that injured nine persons would logically be plotted as a single incident on the control chart, whereas the OSHA log would reflect nine recordable injuries. That is what is actually plotted in Figure 4.7.

At month 16, group 1 seems to have found ways to reduce the wild variation experienced in the preceding months. In fact, we might be concerned about the data. A run of 15 consecutive points within 1 standard deviation of the mean is an out-of-control event and suggests that the data are being manipulated. This criterion is not reached here, because of the low number at month 24. At month 16, there is a process shift that runs for another ten months, followed by a second process shift. A review of group 1's efforts reveals a steady application of actions aimed at changing the safety process and culture of the group. After the second shift occurs at month 26, we see an apparent downward trend through month 40. Nevertheless, sticking to the rules of charting, group 1 resists the temptation to claim victory again when months 35 through 40 are below the mean (six of the seven required). Group 1 remains confident their work will yield a true shift in the future, even though month 41 fell slightly above the mean.

Group 2's results are shown in Figure 4.8. Several new points can be made by looking at this chart. First the upper control limit (UCL) and lower control limit (LCL) are much wider than was the case with either of the preceding charts. Since the control limits are proportional to the process mean and inversely proportional to the group size, group 2, with a higher mean and smaller group, has wider control limits.

Group 2 is much smaller than the groups in the two preceding charts. For the Figure 4.6 the employee base used is 6,000. For the group 1 in Figure 4.7, we have 4,000 employees. Group 2 in Figure 4.8 has 1,800 employees.

We can see an upward shift around months 23 to 28. This shift illustrates a process shift in the wrong direction, with 7 consecutive points above the previous mean. On closer examination, the next series of points are very close to the new mean. A special cause requires finding 15 points within 1

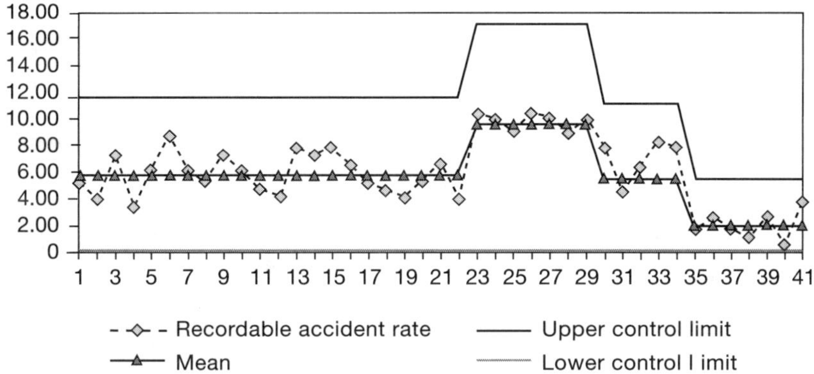

Figure 4.8 Accident rate for group 2.

standard deviation of the mean. In this case, this condition is not met, as there is another process shift at month 30. However, the points between months 23 and 29 are much closer to the mean than 1 standard deviation, leading us to suspect the presence of a special cause. Upon investigation, we find that because of the upward shift in the incident rate, managers in group 2 put extreme pressure on the group to hold down the accident rates. In their zeal to turn the trend, they did not realize they had gotten specifically what they had asked for. People stopped reporting accidents. They kept having accidents at the rate the process would have been expected to produce them, but they simply did not report those they could hide or classify as not being recordable. As the number of accidents in the month increased, the pressure not to report increased. Part of the reason the rates did not drop further was that some accidents were just too serious to hide or underreport. The existence of nonrandom patterns frequently indicates that someone is manipulating the data.

In fact, the control-chart-interpretation rules are designed to identify nonrandom conditions in the data. No set of rules can cover all such conditions. If there is a clear and repeating pattern in the data, you are likely to have a special cause present. For example, if points consistently alternate, one above the mean and one below the mean, there is likely a special cause.

When they realized what was happening, they immediately took action to prevent accidents rather than just avoid counting them. The initial result was a return to the previous process mean. Then, after continued focusing on accident prevention, they experienced another change in the right direction. We never found the cause of the upward shift in the process.

Moving on to our group 3 chart in Figure 4.9, we can see a different pattern. The symptom of reducing group size is even more evident here.

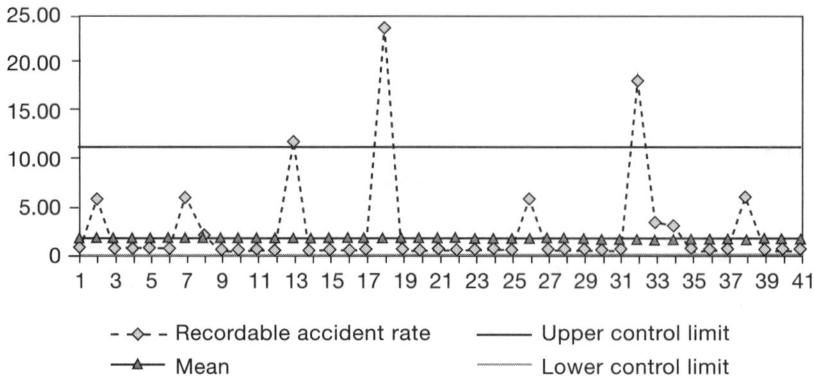

Figure 4.9 Accident rate for group 3.

Group 3 has only 200 employees, and the effect is evident. The LCL is always at zero, and the UCL is very high, considering the low incident rate. The process appears to be in control except for three points at months 13, 18, and 32. On investigation of the out-of-control points, no special cause can be found. The accidents are actually quite routine and common for the group. One problem with this *U* chart is that the employee group is small. However, the out-of-control point at month 18 represents four accidents for the month in a site that averages two accidents per year. This is a major change. Our experience with small plants is that this is not uncommon. The plant may go many months without an incident and then have several in the same month. It may be that the small size of the plant leads to some group reaction to an incident that triggers others. In a larger plant, where there is more social distance between employee groups, this may be less of a factor. This is only a hypothesis, but is the best one we have.

The *U* chart is a critical tool for avoiding the overmanagement of a system that is in statistical control. It has been our experience that the *U* chart is frequently successful in helping traditional managers understand that accidents are produced by the process not by incompetent or unmotivated employees. These managers can then begin to look at safety improvement with a longer-term view and a systems perspective.

The *U* chart helps us recognize accidents for what they are. They are the result of the design and operation of the process producing them. The further back in the process we can move to intervene and interrupt the sequence of complex events that result in the accident, the better our chances of preventing not only the accident types we are aware of and have experienced in the past, but also those new accidents we might never have predicted and, therefore, never have taken action to prevent.

Pareto Analysis of Incident Information

Pareto's law is what we know in common language as the 80/20 rule, which is that most of the incidents come from a small set of causes. Finding those causes and eliminating them is a major source of leverage. Of course, this works only if incident investigations effectively find upstream, systemic causes. If 50 percent of investigations find operator error to be a primary cause, then it is better to improve the investigation process than to attempt a Pareto analysis. We will discuss this further in Chapter 5.

KEY POINTS

- Incident-based measures are the most widely employed, and recording incidents is required by OSHA.
- There are potential reliability issues because of variation in the criteria for determining whether an incident should be recorded under OSHA regulations.
- Additional reliability problems are posed by incentives for workers and managers, rewarding them for low recordable-incident rates. This may cause nonreporting, or manipulating criteria so that reported events are determined to be not recordable.
- Because of the recording criteria, companies may focus considerable resources on preventing trivial events, which must be recorded. At the same time, serious events, such as a chemical process going temporarily out of control, for which recording is not required, may be neither recorded nor investigated.
- In refineries, incident rates are not a good predictor of major process incidents, such as fires and explosions. Releases are a much better predictor.
- These measures, with their inherent variation, especially with small populations, are likely to generate tampering, namely, management intervention based on the assumption that the process has shifted when it has not. To avoid this, incident rates should be plotted on a control chart, using the U statistic.
- Close inspection of the control chart can reveal shifts in the process of accident prevention and can also reveal manipulation of the recording process.

- As an end-of-process measure, incident-based measures are of limited usefulness in understanding the systemic causes of incidents and, therefore, have a limited usefulness in preventing them.

ENDNOTES

1. The only accidents that do not have a human element are those that in no way involve humans. For example lightning striking a tree and causing a fire could be an accident with no human element. Unless of course the tree was planted by a person or someone cut down all of the surrounding trees leaving this one as the highest most vulnerable to lightning strikes. For the most part, only pure acts of nature are the only accidents that can be said have no human element. Even then the consequences are frequently affected by where humans build, remove, or otherwise alter the natural course of events. A tidal wave is an act of nature. The destruction of a coastal city has a human element because humans built there in the first place. Earthquakes are an act of nature. The degree of damage is greatly dependent on where people built and the way they built their structures.
2. I wish I could recall the speaker at that conference to give him credit for the spark of this idea, but his name is lost to me. This just goes to show the oftentimes remote and distant effect of ideas thrown to the masses. Could this be analogous to the chaos description of a butterfly flapping its wings in Asia and causing a tornado in Kansas?

5

Incident Investigation

Incident investigation is a critical component of the process of using incident-based measures. Without an effective investigation into causes, recording incidents does nothing to prevent them in the future. Viewing investigation as part of this measurement process brings with it the realization that its reliability and validity are important. If this process is not reliable and valid, it may lead us to take the wrong actions.

The particular approach one takes to incident investigation depends in part on one's theory of incident causation. Investigations based on Heinrich's domino theory would seek to find the culprit or circumstance that tipped the first domino. In our experience, a culprit can usually found. The majority of investigations that we have reviewed point to an error, usually an error made by an operator. The prevalence of such findings often leads managers to assert that if they could just get better workers, there would be fewer accidents. These findings, in turn, lead them to create motivational programs, on the assumption that such programs will lead workers to work more safely.

A better approach would be to understand the systemic causes of accidents. Deming asserted that at least 85 percent of defects, including accidents, derive from the system not from individuals. Adopting a method of incident investigation that offers an opportunity to view the system can get us out of the trap of focusing on the supposed transgressions of hourly workers. However, finding causes in the system should not be simply a way to shift blame from workers to managers. Blame fixing is not productive in either case. When we understand systemic causes, we are usually in a better position to drive the implementation of system changes.

In our experience, methods like the causal-tree analysis, management oversight and risk tree (MORT), and the fishbone diagram are all suitable for understanding the systemic causes of an accident if they are managed appropriately. However, the devil is in the details here. It takes considerable knowledge, skill, and experience to manage these processes effectively. In our experience, the particular method is less critical than is the ability to truly understand what a systemic cause is and the ability to facilitate the group to ensure open and active participation by all.

The objective of incident investigation, as with any other form of safety measurement, is to guide actions to make the operation safer. Finding the particular causes of a particular accident will likely do little to accomplish this broader objective. We may prevent this particular incident from recurring, but we may actually increase the likelihood of other accidents, by diverting resources and attention to the particular aspect of the single incident we just had. If our home is entered by a thief who slipped the bolt on the front door, we are probably well advised to install a deadbolt. But it would be a dangerous illusion to think that we had made the neighborhood safer.

What we should be looking for are system causes that are common to many incidents. Dealing with these causes effectively requires even more creative effort than the process of discovering them, but this is where the real leverage for improvement is.

RELIABILITY

We are not aware of a single study that has inquired into the reliability of incident investigation. What would be required is the investigation of a series of incidents by at least two independent teams and a comparison of the findings. We do have informal data on reliability. One of us has trained over 100 groups in the causal-tree investigation process. The final step in the training is to break the group into two or three teams and provide them with the same set of facts. Our estimate is that over 90 percent of the teams develop the same basic logic diagram. This convinces us that the process can be made reliable.

Given the fact that incident investigation is a process at least as complex as auditing, and given the very limited reliability of auditing (see Chapter 7), we believe that extreme care is necessary in order to ensure a reliable investigation process. In order to maximize reliability, it is necessary to have a very clearly defined process. This definition should include who should investigate and how.

Who Should Investigate

A person experienced in root-cause analysis and skilled in facilitating groups should lead the investigation. In our opinion, specific expertise in safety is helpful but not required. Other participants can bring this to the group. The group should include persons who are directly involved in the work process where the incident occurred. The group should be structured to maximize the amount of information uncovered. People who will stifle this should be left out, even if they hold important positions. The group should be flexible in its membership. If the investigation points in a direction where more information is needed, the relevant individuals should be invited. These groups generally function best if they include six to ten people. Smaller groups tend to be more easily dominated by one or two persons. Larger groups will work only if the facilitator is sufficiently skilled.

The role of discipline. One of the major barriers to open sharing of information in accident investigations is the expectation of participants that discipline will be assessed. One company we worked with went so far as to declare a moratorium on discipline for anything discovered as a result of an accident investigation. This action may seem like an extreme measure, but when the site considered the reality of disciplining as a result of the findings of accident investigations, they found two disturbing facts.

First, when an employee was disciplined after an accident violating a safety rule, in almost every case that same employee had done the same things many times before having the accident. More often than not, the employee's coworkers and supervisor had actually seen the activity being done and not taken action to stop the employee or others from doing it. The reality was that the injured employee was being disciplined for getting caught (having the accident), not for the safety rule violation, as we thought.

Second, the vast majority of the disciplinary actions were directed toward nonmanagement employees. Very few were exercised against supervisors and even fewer against managers. If we truly believed that accidents are caused by the system, how could these managers have been ignored as contributors to the accident? The answer was not so much that supervision and management were shielding themselves as it was that we were simply not looking deeply enough into the process. When the investigation found "that darn employee violated the safety rule," we tended to stop the investigation, because we thought we had found the cause.

By removing blame as a cause, we hoped to get to the more systematic causes, including such things as management involvement, training, job design, and other system factors. In our discussions of this idea of a moratorium on discipline for involvement in an accident, we considered the more fundamental concept of discipline as a corrective action. Much to our group's surprise, we began to question many cases in which we had administered discipline. At one point, one of the team members pointed out that sometimes you must exercise discipline just to show rule violators the consequences of their actions. He went on to say, "That is how people learn. That is how we teach our kids, isn't it?" To which another team member simply asked, "When was the last time you disciplined your wife?" The point was vividly made that we had been treating adults like children, and it was only natural that the employees' response was more one of rebellion than change. The discussion about suspending discipline for accident investigation findings was over. We decided to try the moratorium.

The issue with discipline is partially fear but also fairness and cooperation. To have open and honest communication in the accident-investigation process, it must not be viewed as an inquisition to assign blame. This was an example of how a good tool can be misused. At the site that had the problem, the causal-tree method had been nicknamed the "hanging tree" by the nonmanagement employees. Obviously, we were not getting good accident investigations, and something fundamental had to change.

How to Investigate

The organization should use a consistent process designed to create a system-wide view. In the Center for Chemical Process Safety book *Guidelines for Investigating Chemical Process Incidents* (CCPS 1992), the authors (one of whom was Pat Ragan) discussed three fundamental approaches to process safety-incident investigations:

> Type 1: Traditional, informal investigation performed by immediate supervisor
>
> Type 2: Committee-based investigation using expert judgment to find a credible solution of cause and remedy
>
> Type 3: Multiple-cause, systems-oriented investigation that focuses on root-cause determination, integrated with an overall process safety-management program

In the update of this book (CCPS 2001), they discuss again the idea of multiple-cause, system-oriented investigation methods as being more rigorous and therefore the preferred method for serious-incident investigations. Either of these books is a good source for details of how to conduct a good incident investigation. It is not our intention to repeat that level of detail here. We both agree that a good logic-driven, systems-oriented method that leads investigators to fundamental system causes of the accident is essential to achieve a system of continuous improvement in accident prevention.

The investigation-system style we prefer is one that uses free-flow logic that documents the relationships among the facts of an event. These are usually referred to as *causal-tree diagrams*. We prefer strong logic theory and a simple process because, in our experience, they lead to consistent results and important insights into fundamental causes. They also have great flexibility in application, and the required training is not difficult. The particular process described here has been used for accident investigations by such major companies as ICI, Rhône-Poulenc, Rohm and Haas, and others.

Causal tree methods begin with listing facts, followed by constructing a diagram (tree) of the facts following specific logic steps or queries. The method we believe strikes the best balance between logical rigor and user friendliness is the logic used in the causal-tree system. It uses the logic tests of necessary and sufficient conditions. The facilitator leads an investigation team to establish a list of facts related to the incident. The team then identifies the end result, the accident, and asks, which of the facts were directly *necessary* for the end result to have occurred? Then they ask whether that fact was, in and of itself, *sufficient* to have caused the end result? If the answer to the second question is yes, that fact then becomes the next end result to be analyzed. If the answer is no, that fact was not by itself sufficient to have caused the end result. They then look for facts that would, when combined with other facts, have been sufficient to have caused the end result. A segment of the tree is diagrammed in Figure 5.1. When the logic test answers yes to the question of sufficiency, the team then moves on to each of these sufficient facts to analyze them as end results. The refinement of the tool by Rhône-Poulenc adds further to the system perspective by expecting each investigation to identify an actionable target in each of three categories: human, organizational, and material. This is intended to add redundancy and diversity to the actions and increase the likelihood of effective intervention.

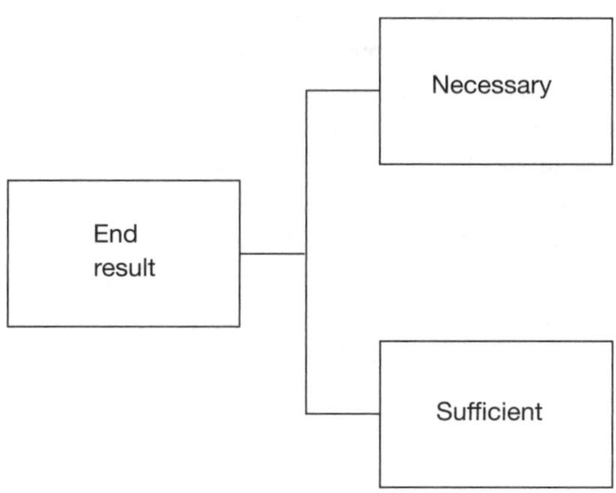

Figure 5.1 A segment of a causal tree.

In the example depicted in the logic tree in Figure 5.2, a site had a complete failure of the bottom of a fiberglass product tank and a resulting spill of product. The logic dialogue would have been:

- Regardless of why the tank bottom failed, the release would have happened. Therefore, the tank-bottom failure was both necessary and sufficient to have caused the product spill. The facilitator stops discussion on this level and asks the team to consider why the tank bottom failed.
- If the tank was overpressurized,[1] regardless of the reason, the tank bottom would have failed. Again, both necessary and sufficient tests have been passed, so the team moves on to the next level.
- If the tank had been overfilled but the relief valve had worked properly, the tank bottom would not have ruptured, and the relieved pressure would have been diverted to a safe location. Therefore, overfilling by itself was not sufficient to have caused the tank-bottom failure. The facilitator asks the team to identify other causes. They then add the failure of the relief valve. If both overfilling and valve failure occurred at the same time, the result would be over-pressurization, which then would have resulted in the tank-bottom failure. Together these two facts were necessary and sufficient to have caused the tank-bottom failure. The facilitator stops discussion on this level, then leads the team to each of the new end points.

This process continues through the tree to rational stopping points or to a point where the team runs out of facts. The team should know that in almost every case there will be leftover facts. In the fact-collection process, it is critical to cast a broad net and understand that not all of those facts will have been factors involved in this specific incident. It is also important to understand that through the logic process, the team knows which facts are relevant to the incident and which are not. For example, in this case, the ambient temperature might have been appropriately identified as a fact early in the process but later found not to have had an effect on the incident. In this case, it is appropriate to have identified and discarded that fact. Conversely, in most cases, not all necessary facts are identified in the initial fact-finding work. This is not only normal, it is a strength of this process. Where facts are missing, the logic drives the team to recognize that they cannot answer the sufficiency question on a branch and must go look for additional facts to consider.

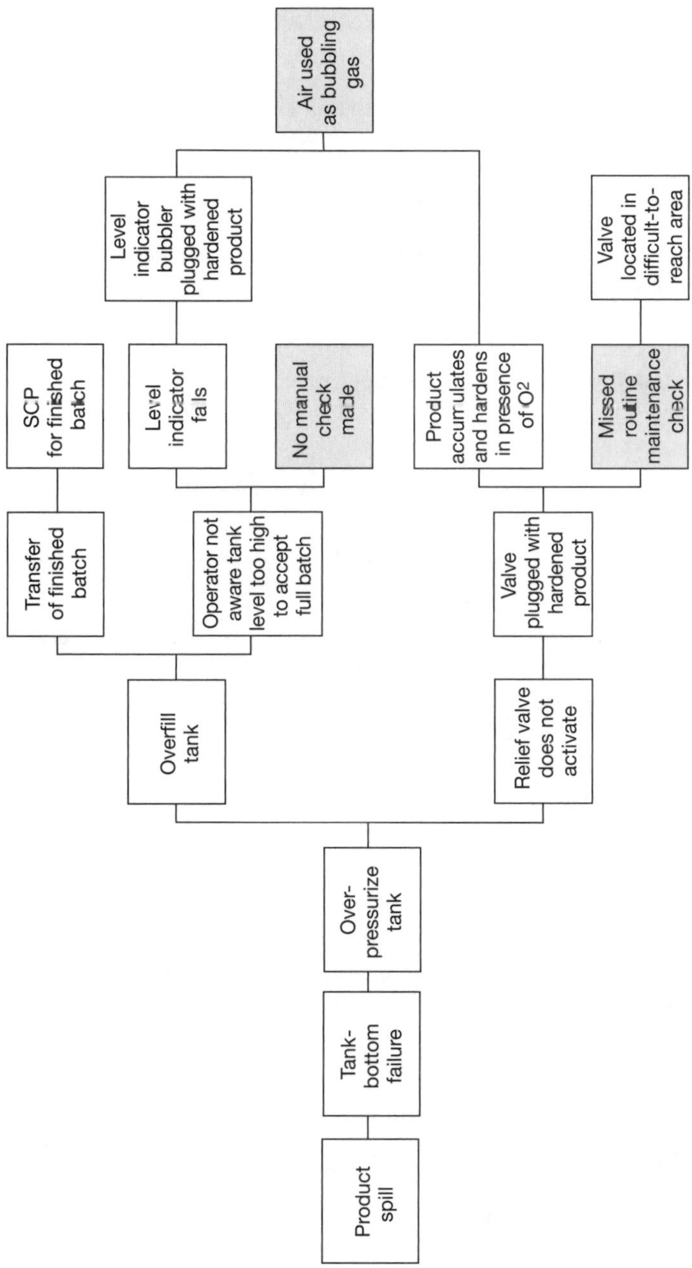

Figure 5.2 Logic tree for the tank-bottom failure.

The description of the incident can be read from a well-constructed logic diagram, depicted in Figure 5.2. In this example the reading of the tree might be:

- A product spill occurred when the process-tank bottom failed. The tank bottom failed because the tank was overpressurized. The tank was over pressurized because it was overfilled and the relief valve did not activate and relieve the pressure.
- The tank was overfilled because a completed batch was transferred into a tank the operator did not know was too full to accept the new batch. The batch was transferred because the batch step was completed and the standard operating procedure (SOP) called for the transfer. The operator did not know the tank was too full to accept the batch, because the level indicator failed and there were no other indications of tank level.
- The level indicator failed because it was a bubbler type, using air in a process in which the material being measured was reactive with oxygen in a way that caused it to build up and harden where contacted by oxygen (air).
- The tank was overpressurized because the installed relief valve did not relieve the pressure caused by the overfilling. The valve failed because it was found to have a buildup of hardened product on the inside surfaces, preventing it from relieving at the designed pressure. The buildup was on the interior surfaces of the relief valve because the oxygen was introduced into the tank via the level gage that was the bubbler type, using air as the bubbling gas.
- Routine testing of the relief valve was on a frequency that would have found the buildup before it was thick enough to interfere with the operation of the valve, but the last scheduled check had been missed. The check was missed because the valve was in a difficult-to-reach location. The maintenance employee put it off for two cycles, which allowed the buildup to thicken to a point where it interfered with the valve operation.

When the logic diagram is followed, the description is accurate and efficient. This approach also helps keep out of the description the opinions and excessive explanations that only tend to cloud the factual nature of the description. Such extra text can cause misunderstandings on later reviews and legal problems if the investigation should ever be called into court.

The next step is to identify the three target factors as depicted in Figure 5.3. A general rule of thumb is that the further down the wide part of the tree that the team targets, the better. Generally, the facts influence more of the other facts as they get closer to the wide part of the tree. However, a good, strong actionable item anywhere on the diagram is always a good choice. For example, if this product were not affected by air and had a very low vapor pressure with no pollution or toxic characteristics, it might be possible to remove the relief valve altogether and just have an open gooseneck vent that could not get plugged. Because this is not possible in this case, the team elected to focus on a *material target* of the "air used as bubbler gas" fact. After review, air was switched to nitrogen, and the plugging problem was eliminated.[2] The *human target* selected was the "missed routine maintenance check" fact. Because the maintenance employee was well aware of the importance of sticking to critical safety checks, such as a relief-valve check, discipline was administered.[3] However, because it was recognized that the valve was difficult to get to and presented ergonomic hazards, a new access platform was added to make the valve easier to access. In this case, the location might have been selected as an organizational target; but, even with a poor location, the maintenance employee was expected to report the problem and get it corrected, not simply avoid the difficult work. The *organizational target* selected was the "no manual check made" of the tank level prior to transferring the next batch. The reason this was considered organizational was that, on reflection, it was felt one level indicator on this tank was not adequate from a safety-reliability perspective. The process of reviewing the tank for adequate safety controls was found to be defective in that it did not identify this during the review. To address this failure, the review-process rules were rewritten to assure all tanks in batch-transfer mode of hazardous chemical had at least two independent means of level control and interlocks.

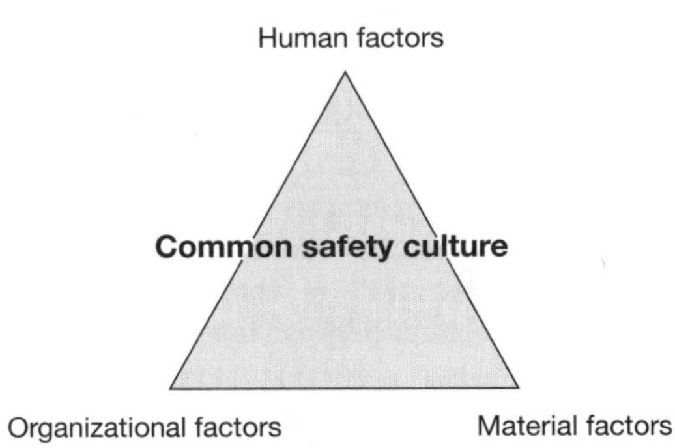

Figure 5.3 Factors in accident causation.

Advantages of the causal-tree method:

- The logic diagram is simple but powerful.
- The process can be understood and used by most investigators without losing the integrity of a strong logic process.
- This method is simple yet robust enough to be used to analyze simple, minor incidents and highly complex, serious incidents.
- The requirement to look for human, organizational, and material targets helps diffuse the tension caused by fixing blame on one area or another.
- The process helps people understand that the causes of accidents are in the system, and the prevention of accidents is best accomplished by improving the system.
- The method tends to cause the investigation team to identify a diverse set of targets and actions that greatly improve the likelihood of finding system causes that will prevent a range of potential accidents.
- The use of cooperative team thinking in identifying facts and constructing logic diagrams results in understanding and commitment by those involved.

Disadvantages of the causal-tree method:

- The method requires participants to think about accident causes in a way that is most likely new to them.
- It sometimes makes managers, workers, and unions uncomfortable, by pointing out that they are part of the system. This can cause them to believe they will be disciplined.

VALIDITY

A well-conducted incident investigation has content-based validity. However, it is difficult to assess the other two types of validity. For criterion-based validity, we are not aware of what might be used as a criterion. We might look at the correlation between system causes of quality problems, as discovered by process-improvement teams, and the system causes of incidents. We would expect to find considerable overlap. If we failed to find this overlap, we might question the validity of either or both classes of investigations.

Construct-based validity might be demonstrated by showing that investigating incidents and taking action on the findings reduce incident rates. To test this experimentally is virtually impossible, because it would require some sites to refrain from investigating accidents, to serve as control cases. An alternative would be to evaluate the quality of incident investigations at a number of sites with similar risks and then look at the correlation between investigation quality and incident rates. Better investigations should generate lower accident rates. Of course, it might be that well-managed sites have better investigations and lower rates, but these two factors are not related. The second part of the test would be to improve the quality of the investigations at the sites that were not doing them well and see if incident rates declined. This would imply a causal link. Although it seems logical that effective accident investigations should improve performance, we can find no demonstrations of this in the published literature. We do have an unpublished case we feel supports the argument for construct validity of quality incident investigations as a positive intervention method for improving safety performance. This case is described next.

Safety Practices in Israeli Aviation

In 1994 I was attending a conference on risk management hosted by the Air Force Safety Agency (now called the Air Force Safety Center). Senior representatives of all branches of the U.S. military attended the conference. While there, I met a colonel in the Israeli Air Force who told the following story:

> *In a single year in the early 1980s the Israeli Air Force had about 30 class A accidents during training. (A class A accident involves the loss of a pilot, an aircraft, or damage in the amount of at least $1 million.) The Knesset (Israeli Parliament) called in the commander of the Air Force and told him that the country could not afford this and they had to do much better.*
>
> *Part of the response was to set a policy under which all incidents, including near collisions, and so on, were reported and thoroughly investigated as to the root cause. I believe that the colonel told me that Simon Goldrat, a well-known Israeli quality expert, was involved with the Safety Center of the Israeli Air Force. Over a period of several years, through the finding and eliminating root causes, the incidence of class A accidents was reduced to near zero.*

We could find no written documentation of this, either in journals or through a comprehensive search of the Internet. Consequently, we called the Air Force Safety Center, currently located at Kirkland Air Force Base, and spoke with the director. He told us that our story was essentially true and referred us to Mr. Itzhak Raz, in Israel. Mr. Raz is currently director general of the Israel Civil Aviation Administration. Since graduating as a combat pilot in 1972, he has investigated and served as a chief investigator on over 600 investigating committees regarding aircraft accidents, both military and civil, and he is head of the Aviation Safety and Security Research Center, funded by the Israel Air Force Association. He is also an attorney, having studied law after his departure from the Israeli Air Force. He was very generous with his time and his immense expertise, spending several hours with us in telephone conversation and reviews of our written account.

According to Mr. Raz, the annual rate of class A training accidents in the Israeli Air Force was reduced from above 35 per year, during the 1970s and 1980s to fewer than 3 per year in the late 1990s. This low rate still continues today.

Three years ago Mr. Raz became chief investigator of the Ministry of Transport. Using his experience from the Air Force, and in collaboration with the knowledge and expertise of Israel's second investigator, Uri Dayyan, they trained more than 250 persons to be investigators. These included pilots, captains, flight attendants, and mechanical technicians from all over the country. They were involved with all types of aircraft, from Boeing 747s to small gliders or parasails. Over the last two years, these trained investigators have served as volunteers in support of Mr. Raz and Mr. Dayyan. Together they increased the number of investigations by a factor of six in comparison to previous years. The results were immediate. The rate of fatal accidents was reduced from an annual range of 8–11 to a single accident over the last two years.

In reviewing the history of this process, beginning with the Israeli Air Force, Mr. Raz told me that one of the first steps the Air Force commander took was to expand the flight-safety branch to a department that includes five branches:

- Investigation
- Flight safety
- Research and data
- Ground safety
- Quality inspection

The investigation branch is most important. In the past, officers who were sent to the investigation department were of average quality. Today only the highest-rated officers are sent. Previously, only three years of experience were required to serve in this branch, whereas now the job requires six to ten years of experience. During their service, officers in the investigation department are usually trained in the United States, and they participate in many conferences. They eventually become essential participants in carrying on the culture of the Israeli Aviation Safety Center and Quality Inspection Department (ASQID).

The investigation process they use does not conform to a detailed protocol or script. The principles are critical, however. According to Mr. Raz, the key principles are:

1. The process must be directed by a skilled and experienced investigator. This person is basically the chairman of the investigation board. He or she is not just a facilitator but an active participant in the process. As already mentioned, these investigators are highly rated officers with at least six years of experience.

2. The investigation is carried out in a step-by-step manner. The last question to be addressed is *why* the accident happened. Considerable care is exercised in laying out the facts of *what* actually did happen, and then *how* it happened. All areas of inquiry should be opened. Finally, recommendations for action are made. These must be clearly structured so they can be accepted or rejected by the commanding officer. In the past, an investigation typically took more than nine months. The current goal is one to two months.

The objective is to find recurring system causes of incidents. Therefore, no single investigation is critical. Finding such causes involves investigating a series of accidents and then constructing a Pareto analysis of the causes. Taking action to eliminate the most frequent causes provides leverage to prevent many accidents not just those that are very similar to the one you just investigated.

The more incidents reported and investigated, the more likely it was that system problems could be found before they led to an accident. As class A accidents became rare, investigators began to investigate other incidents, such as near collisions. According to Mr. Raz, it does not matter what event you investigate. A thorough investigation process will usually uncover the same system causes. They even simulate incidents and investigate the response. Again, the same causes are likely to appear. The causes lie in the system. They surface as problems, causing either accidents or incidents. By investigating incidents, they can discover and eliminate hazards before they cause accidents.

Many times the head of ASQID suggested the Air Force commander stop flights for a day and use this day for safety instruction. For example, if there were many near misses in the air on the last warm days before winter, a day was taken to discuss these incidents.

The Israeli Air Force has adopted a system for reporting every incident that occurs during training flights. So long as an incident is reported within 24 hours of the mission, there is no discipline and no shame. This reporting is required of every officer who flies. Mr. Raz told us a very funny story about a visitor from a foreign air force who told him that in his country such reports were required only for the rank lieutenant colonel and below. Mr. Raz pulled out a file and showed the visitor a report. It was an incident in which the commanding general of the Israeli Air Force had struck his wing on the hangar while taxiing. According to Mr. Raz, that general had a problem with knowing the location of his wingtip, having had three similar incidents earlier in his career.

A current practice is to make video recordings of every pilot who reports an incident. The latest video is circulated to squadron commanders on a monthly basis. To give an example of the detail the process entails, it turns out that bird strikes were determined to be a frequent cause of incidents and accidents. The Israeli Air Force spent two years charting the migratory patterns of birds in their region, using pilots in gliders. The detailed charts they produced enabled them to reduce the frequency of bird strikes by a factor of ten.

According to Mr. Raz, investigation is only a tool not a goal. The goal is flight safety based on accurate data and based on the focus that safety is equally as important as operational efficiency. Every two to three years ASQID introduces a new method to the Air Force including reporting, risk management, total quality management (TQM), and monthly videotapes about news and safety. Because everyone is busy in operations and safety, it is very important to have a research branch and to continue long-term programs for research.

However, in Mr. Raz's opinion, the investigation process is not the most important part of the system. There are two critical changes that must be made first:

1. Leadership must understand that safety and quality are one in the same. There was formerly a belief in the Israeli Air Force, and in the U.S. Air Force as well, that taking risks in training was necessary to build excellence in combat. The assumption was that operational excellence and safety were in competition. In fact, the Israelis have established that exactly the opposite is true. Safety and operational excellence work together. Careful attention to

proper methods in training builds better combat pilots. This is equivalent to industrial managers learning that safety, quality, and productivity go hand in hand. So long as the culture believes that you have to take unnecessary chances to become an excellent pilot, then recommendations to make the process safer will be ignored.

2. The Air Force was the first to require that the chief safety officer of a squadron be the commander of the squadron and not a pilot or an officer. All the responsibility for safety is on the commander. He or she can use a safety officer as an advisor or consultant but never in the chain of command. The Air Force commander emphasizes safety in every form. Safety is discussed in every briefing.

Only by establishing these cultural norms can we expect to see the full benefit of recommendations that come from incidents.

The Israeli Air Force experience is an elegant demonstration of the power of the PDSA cycle. Much of the critical information for planning comes from the investigation of accidents and incidents. Repeating the cycle again and again has reduced incidents to near zero.

Figure 5.4 was supplied by Mr. Raz. It depicts the decline in what they now call Class 5 accidents, which involve the loss of an aircraft or a pilot, or damage exceeding $1 million. It depicts the decline from over 30 incidents per year in 1980 to a level of zero in 2001 and 2002. Mr. Raz has asked that we not publish numbers, but rather focus on the relative change, which is quite apparent.

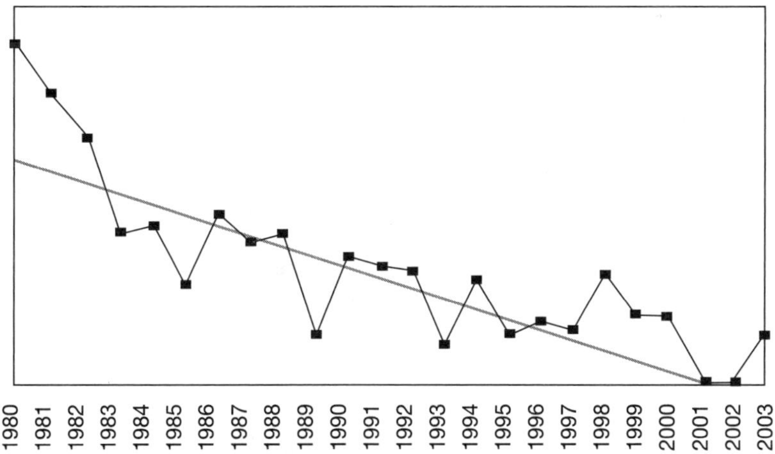

Figure 5.4 Decline in Class 5 accidents.

KEY POINTS

- Incident investigation is an important source of information for system improvement.
- Proper incident investigation requires a well-trained process leader.
- It is critical to cast a very wide net in considering sources of causation. We have successfully used the categories of human, material, and organizational and required the listing of at least one cause under each heading.
- Fixing blame or administering discipline can lead individuals to conceal important information.
- It is important to conduct a series of investigations and look for system causes that repeat. If there are very few accidents, then near-miss incidents or what-if drills should be investigated. These will likely reveal system causes and enable the prevention of future accidents.
- The experiences of Israeli aviation show that effective investigation, coupled with the appropriate cultural norms so that fixes can be implemented, can drive accidents to near zero, even in a very hazardous environment, such as military flight training.

ENDNOTES

1. It is understood here that overpressurization is pressurization to a point that would have caused the tank bottom to fail.
2. The reader should remember this is a simplified example. In practice, the team would consider the effect of switching to nitrogen in a management-of-change review to be sure the action taken would not introduce new hazards, such as asphyxiation from the use of nitrogen in place of air.
3. This was not in the plant in which the moratorium on discipline was established. In retrospect, we would probably not recommend discipline in this case.

6

Systematic Observation of Behavior

In 1980, Thomas Krause and his associates began the application of *applied behavioral analysis* to industrial safety. Applied behavioral analysis is the application of the behavior principles of B. F. Skinner to practical situations, such as psychotherapy. Initially, Krause's work involved systematically observing behavior in the workplace with the objective of improving safety performance.

THE FIRST STEP IN THE OBSERVATION OF BEHAVIOR IS TO DEVELOP A LIST OF BEHAVIORS THAT ARE CRITICAL TO SAFE WORK.

The process is based on the assumption that unsafe actions are a significant cause of accidents. The first step is to identify *critical behaviors*. These are behaviors that are critical to safety. Krause (1997) lists four ways to identify these behaviors: (1) behavioral analysis of injury reports, (2) worker interviews, (3) observation of the work process, and (4) review of work rules, job safety analysis, procedures, and so on. Krause feels that 1 is most important but is not simple. Documenting incident reports is usually not specific to behavior patterns. The analysis usually requires getting additional information from operators familiar with the actual work practices.

The inquiry should include a root-cause analysis of a number of incidents, looking for patterns. This increases the probability that the behaviors are indeed critical. We would add that most of the hundreds of incident investigations we have seen do not get to root causes. Krause is clearly aware of this and provides an excellent methodology to improve this process. He recommends the causal-tree method. This particular method for incident investigation is described in Chapter 5.

Krause, like Deming, points out the importance of operational definitions as key to any systematic measurement. An *operational definition* of a behavior is a definition that would allow many different observers to make the same decision as to whether the behavior occurred. For example, if a critical behavior is the wearing of eye protection, you might define it as "wearing eye protection whenever a hazard is present." This is not a good operational definition. It does not define eye protection. Are glasses enough? Are safety glasses enough? Are side shields required? Nor does it define what it means for a hazard to be present. A better definition would be, "wears safety glasses with side shields whenever present on the factory floor." This is not necessarily the optimal safety rule, but it would allow observers to judge with good reliability whether it is followed.

Starting with the critical-behaviors list, operational definitions are created for safe and at-risk versions of the behavior. At-risk behaviors include failure to wear personal protective equipment and standing in the line of fire while conducting an operation. An example of being in the line of fire is standing in the fall pattern of objects, such as a load being hoisted overhead. If something goes wrong, you are in the line of fire and are in danger of being struck by the falling load.

Initially, a baseline measurement is made of the relative frequency of the safe and at-risk versions of the critical behaviors. Then, feedback is provided from an observer to the workers observed, in order to increase the frequency of safe behaviors and reduce the frequency of unsafe ones. Typically, this is not done by a safety therapist, but by an employee who has participated in a special training program on behavioral observation. Krause and others have done a large number of studies to show that this intervention method dramatically decreases incident rates and that decline in incident rates corresponds in time to the decrease in the proportion of unsafe behaviors.

It is important to note that Krause is very careful to point out that observation of behavior is not even the principal component of his behavior-based safety-improvement process. According to Krause (2003), behavior-based safety has "much more to do with understanding organizational functioning, proper implementation of organizational change initiatives, organizational culture, proper use of statistical analysis, and organizational development." However, it is clear that behavioral observation is a widely used method of measurement, and it is appropriate to discuss it in some detail.

RELIABILITY

WITH PROPER ATTENTION TO OPERATIONAL DEFINITIONS AND TRAINING OF OBSERVERS, THE OBSERVATION CAN BE MADE VERY RELIABLE.

Because this methodology was developed by social scientists, we would expect that reliability would be an important concern. In fact, it is clear from the careful attention to operational definition and the training of observers described by Krause, that behavioral observation can be made very reliable. For example, Williams and Geller (2000) demonstrated a 90 percent agreement between two independent observers in the classification of a variety of behaviors. Although it is clear that such levels of reliability *can* be achieved, it would be prudent for any user of the process to test reliability from time to time to ensure that these levels *are* being achieved. Again, this requires two observers looking at the same behavior and recording it independently.

VALIDITY

Content-Related Validity

THE MANNER IN WHICH THE LIST OF CRITICAL BEHAVIORS IS DEVELOPED ENSURES A DEGREE OF CONTENT-BASED VALIDITY.

The manner in which the list of critical behaviors is developed ensures a degree of content-based validity. These are behaviors that are related to incidents or that safety professionals believe to involve risk. Krause argues that behavior is the final common path for accidents. Although this is easy to demonstrate when people fall, catch their hand in a machine, or are struck by a falling object, the connection is often much more complicated when a plane crashes or a refinery catches fire. There are undoubtedly acts and failures to act involved in each of these, but an analysis of individual behavior is a rather inefficient way to understand the root causes of these events.

Consider the destruction of the space shuttle Columbia. The precipitating event was damage to the wing structure caused by falling insulation foam at launch. By a stretch, one might cite the behavior of design engineers who were responsible for the physical system, or flight managers who failed to inspect the craft for damage prior to the stress of reentry. However, according to the Columbia Accident Investigation Board headed by Admiral Gehman:

> *In our view, the NASA organizational culture had as much to do with this accident as the foam. Organizational culture refers to the basic values, norms, beliefs, and practices that characterize the functioning of an institution. At the most basic level, organizational culture defines the assumptions that employees make as they carry out their work. It is a powerful force that can persist through reorganizations and the change of key personnel. It can be a positive or a negative force.*
>
> (Columbia Accident Investigation Board, Report Volume I, page 177)

The recommendations of the board relate much more to the organizational culture than to any behaviors that might be observed.

Criterion-Related Validity

There can be little doubt that behaviors like failure to wear personal protective equipment increase risk. The most dramatic case is wearing seatbelts. Seatbelts are said to reduce the likelihood of a fatality by 45 percent in head-on crashes. When vehicle occupants use seatbelts, only 1 percent of the belted population is ejected in a crash. Unrestrained vehicle occupants are ejected 22 percent of the time. When totally ejected, 75 percent of these persons are killed (Carmody 2003).

WHEN INCIDENTS AND INJURIES ARE USED AS A CRITERION, BEHAVIORAL OBSERVATION HAS A HIGH DEGREE OF CRITERION-RELATED VALIDITY.

There are a large number of studies showing that when a program is instituted to reduce the frequency of unsafe behaviors, a reduction in incident rate coincides with a reduction in unsafe behaviors. Krause, Seymore, and Sloat (1999) report a meta-analysis of 73 firms that implemented behavior-based safety programs for up to five years. Reductions in accident rates from pre-implementation baselines averaged 29 percent after one year and up to 69 percent by the fifth year. Thus, there is criterion-based validity when incidents are the criterion. We are not aware of any studies that have used process safety incidents or fatalities as a criterion.

Construct-Related Validity

Here is where the issues become quite complex. A construct is part of a theory. Hunger is a construct. It cannot be measured directly. It is inferred by measuring things like the amount of food consumed. The effectiveness of the safety-management system is a very broad construct. Again, it cannot be measured directly but is inferred from measures of loss, and the like. Applied behavioral analysis is derived from the teachings of B. F. Skinner. Skinner (1950) asserted that theories that involved "any explanation of an observed fact which appeals to events taking place somewhere else, at some other level of observation, described in different terms, and measured, if at all, in different dimensions" were unnecessary. He did admit, in a paternalistically tolerant manner, that theorizing would probably continue: "Perhaps to do without theories altogether is a *tour de force* that is too much to expect as a general practice. Theories are fun. But it is possible that the most rapid progress toward an understanding of learning may be made by research that is not designed to test theories." Had he been asked what was measured by behavioral observation, he would probably have answered, "behavior."

Krause does not take this position. In his 1996 book, Krause argues that "Deming has overshadowed Skinner." Whereas Skinner asserted that theories of behavior were scientifically inappropriate, Deming argued that there is no knowledge without theory. According to Deming, the reason for theory is to enable prediction. Although Skinner also sought prediction, he did so in a much more limited sense.

Krause clearly takes a theoretical position. For example, he agrees with Deming that 85 percent of quality problems, including 85 percent of accidents, derive from the management system, leaving a maximum of 15 percent to come from worker-generated problems. Citing the management system as the cause of worker behavior in this way would have been antithetical to Skinner's approach but has allowed Krause and his associates to make considerable progress. The prevalence of unsafe behaviors in a workplace, then, must relate to the management system existing in that workplace. Using this logic, we would suggest that the behavior-based safety process Krause and his colleagues have developed is, in itself, a component of the management system, designed to correct the problems discovered by behavioral observation.

We are aware of no studies that identify the constructs measured by behavioral observation. We can, however, develop some hypotheses. We and a number of other workers have identified four fundamental components of the management system:

1. Management's demonstration of commitment to safety. This is an important dimension of leadership. Do management's actions convey the message that safety is very important.
2. Education and knowledge of the workforce. Are workers properly trained to do their jobs, and do they receive proper safety training? Do they understand their jobs and how to work safely?
3. Effectiveness of the supervisory process. Does the company have standards for work, and are these standards enforced?
4. Employee involvement and commitment. Are employees involved in the planning process, and are they sufficiently committed to caution coworkers about unsafe practices?

All of these factors would ultimately influence the behavior of workers on the factory floor. However, the most immediate and direct influence would come from (3). The supervisory process is related to standards and procedures. Do standards and procedures exist? Are they correct? Do the workers know they exist? Do the workers know what they are? Do the workers actually follow them? The most direct way to answer these questions is by direct behavioral observation. It is possible to find answers by asking employees, but this is limited in several ways. First, if workers are frightened or lack confidence in the person asking the questions, they may conceal the truth. Second, they may not know the truth. Behavioral observation by a trained third party can supply an objective and reliable answer to the questions.

The relationship of the other three factors to observed behavior is less immediate and direct, although training is, in many cases, a direct influence. The commitment of management may have a powerful motivational effect. For example, the perceived commitment of management enters the worker's decision about whether to perform a task the prescribed way or the expedient way.

IMPLEMENTING A PROGRAM OF OBSERVATION AND FEEDBACK REPRESENTS A POWERFUL WAY OF IMPROVING THE SUPERVISORY PROCESS.

Our hypothesis is that direct observation of behavior is most directly related to the quality of the supervisory system. Implementing a program of observation and feedback represents a powerful way of improving that supervisory process, albeit with a cost that may be quite high as well. However, as Krause suggests, observation of behavior alone is not sufficient to drive improvement in safety performance.

This is supported by the observations of Harper and his colleagues (1996). They studied the effectiveness of behavioral observation safety programs at nine work sites across seven industrial companies in Australia. They found that three of the nine programs were successful in reducing the frequency of unsafe behavior and improving housekeeping. Of the remaining six, three were clearly unsuccessful and three were equivocal. In attempting to understand the factors that may have contributed to this, they found that the successful sites, in contrast with the unsuccessful sites, had the following management system characteristics: (1) active managerial involvement in safety; (2) delegation of safety authority to employees; (3) communication about the program through all levels of the organization; (4) consistent management behavior regarding safety; (5) management leadership of improvement hazard detection and improvement efforts; (6) supervisors acting as role models; (7) personable communication between mangers and employees; and (8) a high priority of safe production.

Most of these factors, including 1, 3, 4, 5, 7, and 8, are part of what we have called *management's demonstration of commitment to safety*. This is a powerful driver of safety-program effectiveness. Moreover, the lack of a demonstrated commitment of management would likely undermine the effectiveness of any safety process, no matter what else was done. The implication is that measuring and modifying behavior is a useful addition to a management system in which management's commitment is well established. Although a more skillful application of behavioral technique may have succeeded in all of the companies, it remains that the process works best in an environment created by a good safety-management system.

We have actually employed behavioral-observation methods in a number of plants, although not in the course of a formal study. Our experience is that it was usually of some benefit. Our conclusion was that, in addition to reinforcing standards and procedures, it helped communicate management commitment to the workers. In addition, the interaction between the observer and the worker being observed provided a form of job training.

CONTROVERSY

There is a considerable literature arising from the union movement that is very critical of behavior-based safety programs that focus on unsafe acts by workers (Walker 2003; O'Neill 2002). The general complaint is that these programs are essentially a new method of blaming the worker, as opposed to improving work processes and eliminating hazards. Certainly Krause is very clear that system improvement is more important than correcting worker behavior, since he agrees with Deming that at least 85 percent of defects, including accidents, derive from the system and not from individuals. Moreover, whether or not behavioral observation is appropriate as a sole focus for a safety-improvement program (and we believe it is not), it has no bearing on whether it is a reliable and valid method of measurement.

KEY POINTS

- Behavioral observation is an application of the behavioral methods of B. F. Skinner to safety improvement.
- The first step is to define behaviors that are critical to safe work.
- Trained observers measure the frequency of safe and at-risk versions of critical behaviors in the workforce. Through feedback they attempt to increase the proportion of safe behaviors and decrease the proportion of at-risk behaviors.
- It is clear that systematically observing behavior can be a very reliable measure if appropriate operational definitions are created and observers are properly trained.
- Behavioral observation is a valid measure of safety when the criterion is injury rates. Safety programs that feature systematically observing behavior as an important component have been very successful in preventing injuries.
- For maximum benefit, this measure is best used in conjunction with other improvement methods that impact organizational functioning and culture.

7

Audits

Audits are a very controversial subject. Many people and companies have made careers and fortunes on the assumed value of audits. Surprisingly few have attempted to demonstrate this value scientifically. In this chapter we discuss audits, their reliability, their validity, and their utility in supporting improvement of the safety-management system.

Merriam-Webster's Collegiate Dictionary lists the following definitions for "audit":

> **1 a:** *a formal examination of an organization's or individual's accounts or financial situation* **b:** *the final report of an audit* **2:** *a methodical examination and review*

Obviously, definition 2 would apply to a safety audit. The word audit implies a standard. The purpose of the audit is to find out if the standard is being met. This could be a standard for the number and location of fire extinguishers, whether operators have met certain training requirements, whether the relief valves on pressure vessels have been tested in the time period required, or virtually anything else that management deems essential to safety. Audits can involve any or all of the following: physical inspections, interviews, and review of documents and records.

The concept derives from accounting. A financial audit is a review of the books and records of a company that is conducted to verify the company is following the proper accounting procedures and exercising the necessary financial controls. A financial auditor reviews the books and records of the company, interviews managers, and tests a sample of transactions to provide verification that the transactions recorded in the books actually occurred. In the test of transactions, letters are sent to some of the customers who appear on the company's accounts-receivable list to verify that they do indeed owe the amount stated on the list. A fundamental purpose of the financial audit is to ensure that generally accepted accounting principles (GAAP) were followed in the company's preparation of the books.

Audited financial statements indicated that Enron, WorldCom, and Health South were making substantial profits at a time when they were in fact losing large amounts of money. There have been vociferous complaints about the auditors. At Enron, the auditors were aware of serious problems with the books but did not confront them. This led to the criminal charges against Andersen, the firm responsible for the audit, and to the downfall of that firm. At WorldCom and Health South, management allegedly recorded bogus transactions and lied to the auditors, and the audit methods did not discover this. No action has been taken against these auditors. Although the auditors have been sued by shareholders, they have not been accused of criminal acts.

The failure of auditors to discover some of the most egregious financial frauds in recent years might lead one to question the entire process of financial auditing. However, just as there is an inclination in safety to focus on the people and not the system, the same is true in finance. The problem is blamed on corrupt auditors and greedy executives. Some more-sophisticated complaints target the system in the United States whereby a company can fire an auditor if it does not like the report the auditor gives and find another auditor who will issue a more favorable report. This cannot be done in the United Kingdom, for example. All of these issues may be important, but we would also be advised to question the ability of the auditing process itself to accomplish what it is supposed to accomplish.

To explain this, let us consider the inspection of an automobile, which has a strong analogy to audits. We suggest the following questions: "Are the fan belts tight and not worn?" "Are the brake pads worn past the limit?" "Are the fluids at the proper level?"

No matter how many questions we write, you will think the effort is incomplete. Why do we not just drive the car? That is a critical test. After all, the purpose of the car is to provide transportation. The test with the most validity is driving. Of course, inspections have a place. When you take your car in for service they check (audit) the fluid levels, brake pads, hoses, belts, and so on, to prevent a failure while you are driving. But it is truly possible that all of the things on the checklist could be OK, but the car would still not run or would break down on your next trip.

Back to the analogy to financial audits. It is possible that all the accounting rules are being followed, and that the books appear to show a profit, but the company is running out of money. Although Enron was found to have violated some accounting rules, they could still have reported profits when they were losing money, without those violations.

Designing an inspection for a car is relatively simple. Designing a financial audit for a company is much more complex. Compared to the safety management or financial-management system of a company, a car is a relatively simple system. The audit and the inspection provide a static picture of a dynamic system. The effectiveness of the audit or inspection will depend in part on understanding how the system works. Obviously, we will have a more complete picture of how the system works with a car than with a management system.

An audit or inspection may tell us *if* the system is working in the way we expect, but it will probably not improve our understanding of *how* the system works. We attended five four-day seminars taught by Dr. Deming and do not recall his ever discussing the audit as a method for quality improvement. Of course, he frequently said that we should cease dependence on mass inspection. The word "audit" does not appear in the index of either of his books. To the extent that an audit implies enforcing regulations, he would have questioned it. He probably would have asked what value an audit has in understanding the process.

AUDITS DERIVE THEIR CONTENT FROM THE THEORY OF ACCIDENTS HELD BY THE AUTHOR(S) OF THE AUDIT.

In fact, the basis of an audit is the assumption that its designer does indeed understand the process and has created an instrument to ensure that the process performs properly. Audits then, derive their content from the theory of accidents held by the author(s) of the audit. For example, depending on the author's point of view, the audit might emphasize unsafe conditions, unsafe acts, deficiencies in the safety-management system, or any number of other personally held accident-causation philosophies.

STANDARDIZED AUDITING METHODS

There are a number of widely used and standardized audit systems. Kuusisto (2000) reviews some widely used systems, including a process developed by Diekemper and Spartz (D&S), the Complete Health and Safety Evaluation (CHASE), the safety map, OSHA's Voluntary Protection Program (VPP), and the International Safety Rating System (ISRS), which appears to be the most widely used of all.

All of these have a list of questions or areas of investigation along with a method for scoring each question. Typically, the audit involves interviews with site employees to obtain answers to the questions. The audit protocol lists various specific areas of investigation and describes the criteria for scoring the company's performance on that area. Table 7.1 is a very brief section of a Diekemper and Spartz protocol as used by Kuusisto.

It is apparent from this brief snapshot that this is not a simple process. For example, on level 1, consider the problem of judging whether there is a general understanding of safety. There is no operational definition here. An operational definition would tell us very specifically how to make the judgment. For example, the only way to really know if the understanding was general would be to interview a large number of employees. The operational definition might include "at least 63 percent of employees can describe the company's safety policy." Of course, that does not end the problem. What is the standard for deciding that their description is adequate? Lacking such definitions, there will be variation among auditors and diminished or nonexistent reliability.

A second issue is that both questions potentially have equal weight[1] in the final scoring. Are they equally important? That is an empirical question.

Table 7.1 A Diekemper and Spartz audit protocol.

Activity	Level 1 (poor)	Level 2 (fair)	Level 3 (good)	Level 4 (excellent)
1. Statement of policy, responsibilities assigned.	No statement of safety policy. Responsibilities and accountability not assigned.	A general understanding of safety, responsibilities, and accountability, but not in written form.	Safety policy and responsibilities written and distributed to supervisors.	In addition to the previous items, safety policy is reviewed annually. Responsibility and accountablity are emphasized in supervisory performance evaluations.
2. Direct management involvement.	No measureable activity.	Follow-up on accident problems.	Active direction of safety measures. Management reviews all injury and property damage reports and supervises the corrective measures.	Safety matters are treated the same as other operational parameters (such as quality or production design). Management is personally involved in safety activities.

To answer it one would want to have some data that relates the item score to some independent measure of the safety-management system, such as injury rate or costs. We are not aware of any such published studies.

In fact, there is wide variation in the points of emphasis of the various standardized auditing systems. Kuusisto divides the areas of inquiry into four broad categories: policy, organization, and administration; hazard control and risk analysis; motivation, leadership, and training; and monitoring, statistics, and reporting. Which components are emphasized in the final score depends on the system. Kuusisto describes the differing emphases of three systems in Table 7.2.

These differences are obviously quite substantial. Unfortunately there exists no published research to tell us which affords the most reliable and valid measurement of the safety-management system.

RELIABILITY

There is very little published work on the reliability of audit systems. Kuusisto (2001) conducted an extensive study on the reliability of auditing methods. His dissertation reviews the literature on this topic and reports an original study of the inter-observer reliability of the D&S method and of an "improved" audit method labeled Method for Industrial Safety and Health Activity Assessment (MISHA). In an initial test, Kuusisto audited six companies in the United States, using the D&S method. He compared the audit scores he recorded with the scores of internal audits conducted by a company employee who also used the D&S method. Statistical testing indicated that the reliability was poor to moderate. In only one of the companies did the reliability reach the moderate level. Two companies reached the fair level, two reached the slight level, and one was poor.

In a second test of the D&S method, Kuusisto examined three Finnish companies and compared the scores he recorded with scores independently recorded by his students. Here the reliability ranged from fair to

Table 7.2 Emphasis of inquiry components under various auditing systems.

Category	D&S	CHASE-II	ISRS
Policy, organization, and administration	20%	35%	33%
Hazard control and risk analysis	40%	48%	19%
Motivation, leadership, and training	20%	6%	19%
Monitoring, statistics, and reporting	20%	11%	29%
Total	100%	100%	100%

almost perfect. The company employees in the first test were generally more strict in their interpretation than Kuusisto. In explaining the difference, he points out that the company auditors, because they worked for the company, had much more knowledge about the company than he did, whereas his students were working from the same information he was.

In the study of the MISHA method, Kuusisto compared his scores on a Finnish company with the scores recorded by the company's safety director, safety manager, safety representative, and personnel manager. He found fair agreement between his scores and those recorded by the safety manager and safety director, and only slight agreement between himself and the other two observers.

AUDITS CAN BE MADE RELATIVELY RELIABLE IF CONDUCTED BY INDIVIDUALS WITH SIMILAR TRAINING WHO ARE WORKING FROM THE SAME INFORMATION.

A reasonable conclusion from Kuusisto's studies is that audits can be made relatively reliable if they are conducted by individuals with similar training who are working from the same information. The highest reliability was achieved by comparing Kuusisto with his students, who were working from the same information. An intermediate level of reliability was obtained when comparing Kuusisto's scores with those of other safety professionals, who were employees of the company. The safety professionals had specialized safety training and experience but different information, because they knew much more about the company. If audits are conducted by persons with different training and different information about the company, then reliability is very low. It must be pointed out that this was not a determination of whether the finding of the audit was an accurate representation of the performance of the safety-management system. That is a question of validity. The test of reliability means that the results of the audit are reproducible. The auditors with more experience, training, and personal knowledge (company employees) gave scores that differed more from Kuusisto's scores than did the scores produced by his students. Whose scores provided the most accurate or valid representation of the systems audited is left to question.

Apparently recognizing this problem of reliability, Det Norske Veritas (DNV) offers training for company employees in the conduct of its proprietary ISRS audit. In fact, since 1994, DNV has not maintained a single, standardized audit but has instead worked with individual companies to develop custom audits for the specific needs of that company. Presumably they are able to obtain some degree of reliability as a result of this, although we are not aware of published data on this matter. Overall, the published literature suggests that under the best conditions, audits can be a reliable measure.

VALIDITY

Content-Based Validity

Because audits are typically designed by safety professionals, they would be assumed to have content-based validity. Certainly the questions are those deemed by the designers to be the most important indicators of safety performance. The fact that, at least in some circumstances, reliable scores can be generated should satisfy us that the method has content-based validity. The presence of reliability indicates that several auditors are using similar definitions of the content, which provides a validation of that content.

Criterion-Based Validity

There are very few studies on criterion-based validity of audits. Bailey and Petersen (1989, Polk 1987) describe an attempt to relate safety-program characteristics with accident statistics and monetary losses in a very large study of railroads. They surveyed 18 railroads and scored the them on the following areas of safety programs:

- Safety program content
- Equipment and facilities resources
- Monetary resources
- Reviews, audits, and inspections
- Procedures development, reviews, and modifications
- Corrective actions
- Accident reporting and analysis
- Safety training
- Motivational procedures
- Hazard-control technology
- Safety authority
- Program documentation

Only two program areas had the appropriate correlation with the measures of loss: monetary resources and hazard-control technology. Companies that spent more and employed more-advanced hazard-control technology had fewer and less costly incidents. They report that two areas had counterintuitive correlations (meaning the higher the score, the higher the accident rate): equipment and facilities resources, and reviews, audits, and inspections. Thus the finding is that the more extensive the audits, reviews, and inspections, the less effective the safety program. This could mean that the audits and inspections make matters worse, or it could mean that railroads with frequent incidents and greater monetary losses increased their use of reviews, audits and inspections, perhaps as a remedial measure. Although this is not a direct test of validity, the finding at least suggests that the audits in this study had no criterion-related validity and, more important, that the auditing may have actually impaired safety performance.

Eisner and Leger (1988) reviewed the application of the ISRS audit in South African mines. In fact, the ISRS was initially designed as an audit tool for South African mines. Their conclusion, as stated in the abstract of their study, is that:

> *Despite glowing testimonials, there is no substantially significant evidence that it has reduced injury and fatality rates by improving safety. The ISRS covers some areas of mine safety to a high standard but coverage is not consistent. Certain areas are dealt with only in generalities and others neglected. The practicality of objectively and thoroughly auditing a mine is questioned given the system's long checklists, the great detail required by some questions and the limited number of auditing staff. While ostensibly high standards are set for scoring, more than eighty per cent of gold and coal mines have been awarded four or five stars. Recent disasters have highlighted inadequacies of the system. More generally, there is no correlation between star rating and either fatality or reportable injury rates. A danger of the scheme is the degree of complacency it appears to have engendered. Attempts to reduce South Africa's high underground fatality rates must first come to grips with this.*

These authors, then, suggest a serious problem with the validity of the most widely used standardized audit. Although a failure to find validity is not evidence that the audit is invalid, the lack of positive findings to counter this failure presents a problem for this particular audit.

THERE IS SOME EVIDENCE THAT SUPPORTS THE CRITERION-BASED VALIDITY OF AUDITS.

There is some evidence that supports the criterion-based validity of other audits. Uusitalo and Mattila (1989) found a "fair correlation between audit scores and accident rates in 11 industrial companies." There are other studies described by Uusitalo that show some evidence of validity. Pringle and Brown (1990) report a 12 percent drop in the incident rates among the 2,395 North American companies that used the ISRS in the period 1978–1979. The apparent lack of a control group of similar companies that did not use the ISRS and the fact that the improvement was relatively small leads us to some reservation about the conclusion that the ISRS was responsible for this change. It would not be surprising to find that a sample of similar companies that used methods for safety improvement that did not involve audits showed the same 12 percent reduction in accident rates.

The strongest positive evidence for criterion-based validation comes from a study by Ray et al. (2000) of 25 manufacturing plants in Alabama. They developed an audit on the maintenance function of these plants. The audits were conducted by outside auditors from the University of Alabama. Audit scores were correlated with the recordable rate. Using a rank-order correlation, they found a statistically significant negative correlation between the audit scores and the incident rate, meaning that the better the audit score, the lower the incident rate.

Although the authors call this a pilot study, it is a strenuous test of validity. Because the 25 plants were from different industries and clearly had different levels of risk, there was huge variation in the incident rates. This noise vastly decreases the likelihood of finding the expected correlation. A significant limitation of the study is that it is confined to the maintenance function. Another limitation was that the authors did not test reliability. However, low reliability would create even more noise to interfere with validation. It is important to note that this study was limited to an audit of the maintenance function and cannot be generalized at this point to different audits of different functions.

Construct-Based Validity

We found nothing in the literature regarding what underlying constructs might be measured by audits. Many authors assert that audits are a measure of safety performance. Audits may correlate with safety performance, but they do not represent performance in the sense of a criterion. A measure of safety performance should be related to loss.

EFFECTIVE USE OF AUDITS

In spite of the limitations on both reliability and validity of audits, we believe they are of some use in maintaining an effective safety-management system. Moreover, because they are so entrenched as a tool, we do not expect the use of audits to be discontinued. Therefore, we feel obliged to offer recommendations for how they can be used most effectively. If one plans to use audits, there are several questions that should be asked:

- Who is the audit for? Who is the customer?
- What is the objective of the audit?
- What standards of reliability and validity will be applied?
- Is the design and conduct of a reliable and valid audit worth the effort and resources required?

ERRORS OF OMISSION ARE MORE LIKELY TO BIAS AUDIT RESULTS THAN ARE ERRORS OF COMMISSION.

Who is the customer? In our experience, many audits are justified on the premise that they will help improve the systems being audited and provide information for company management to help them meet their responsibility to maintain safe operation. In this case, the principal customer is management, and the secondary customer is the workforce whose safety is being protected. It is our experience that most audits do not effectively serve either customer. The fact that the audits are used by management to measure the performance of subordinates causes those being audited to put the best possible spin on the answers given to the auditors. We believe that errors of omission are more likely to bias audit results than are errors of commission. If you believe your performance review, and ultimately your promotions and pay, hinge on consistently good audit results, you will do your best to obtain a good score. The financial audits discussed earlier are excellent examples of the same thing. Because the managers answering the questions and the auditors asking the questions have a big stake in the outcome, the results are very likely to be biased.

This bias in safety audits, through errors of omission, is very counterproductive. Although it may be useful and important to identify what is done well, it is much more important to know what is not being done well. If an audit is to facilitate continuous improvement, the auditors need to have a good knowledge of what is working and what is not working to advise the audited group or site what actions and directions should be taken to improve the system. For the audited, these exercises should be about finding causes of system disruption or failures. This leads us to our next question.

We have some anecdotal evidence regarding how setting up an audit to eliminate bias can be more effective. In the 1980s Dow Chemical developed a self-assessment tool for its plants. This was essentially an audit of the management system and included questions about leadership, training, employee involvement, and operating discipline (Carder and Ragan 2003). Apparently, when the audit was initially devised, it was performed by management. The process had no positive effect on the company's safety performance. Subsequently, the audit was performed by employee teams. This audit process then had a very positive effect on the company's performance. The employees, who, unlike the managers, were not being evaluated by the process, were able to make more useful observations.

What is the objective of the audit? This seemingly simple question is inexorably linked to the question of who is the customer for the audit. Is the objective of the audit to determine compliance with established standards? Or is it to understand process or system shortcomings so the owners of the process can improve that process? The first is an audit for management; the second is an audit to improve the process.

IF THE AUDIT IS NOT RELIABLE, IT WOULD BE JUST AS EFFECTIVE AND MUCH CHEAPER TO SIMPLY MAKE UP THE RESULTS.

The audit's objective affects both the form and the content of audit. It is our contention that audits are suited to determine point-in-time compliance. It is relatively easy to design an audit to determine point-in-time compliance with a regulation or a standard. It is much more difficult to design an audit process that will reliably help evaluate and improve a process. It is unlikely that a standardized audit could accomplish this. What is required is that the auditor has an understanding of the particular process under study. Based on this understanding, the auditor can then design questions to evaluate the process. We have some limited experience with this approach. It lacks the appeal of standardization and is relatively slow and costly. Moreover, it can succeed only in an atmosphere of trust. If the auditor is to ever find out what is really happening, the employees involved have to be confident that the information they reveal will be used to improve the system rather than to evaluate them.

To what standards of reliability and validity will the audit be conducted? It makes no sense to conduct an audit of any kind without some mechanism to ensure reliability. If the audit is not reliable, it would be just as effective and much cheaper to simply make up the results. Testing the reliability requires having independent auditors conduct the audit and then statistically comparing the scores.

Many safety professionals believe that their expertise and experience in safety ensure that they will be competent auditors. However, competence is not the issue. If different auditors, in spite of their competence, give substantially different scores, what is the appropriate result for publication? The average? The score produced by the most senior auditor? Reliability in this context is reliability as a measurement tool or a tool to consistently produce safer system results. Audits done by knowledgeable professionals can and do produce improved safety results if actions are implemented. We just have not found a way to predict just how much improvement should be expected or a way to use the results to reliably measure two sites, units, groups, and so on, against one another. Without establishing reliability and validity of the audit process, we cannot use this process to determine who will have the better safety performance in the future.

Audits can lack reliability either because the protocol is defective, because the auditors are not properly trained, or because the employees being audited do not provide candid information. To the extent that the criteria for scoring are not operationally defined, reliability will be reduced.

Validity is another question. We have said that audits typically have content-based validity. Testing for criterion-based validity may be beyond the scope of many situations for which safety professionals would like to apply audits. Certainly in a company with a small population and relatively low accident rates, there would probably not be enough data to find a relationship between audit scores and accident rates. An alternative would be to look for a relationship between audit scores and the scores on our safety survey (see Chapter 8). Since the scores of this survey are extensively validated as measures of system performance, they can serve as a criterion against which to test audits.

IF THE AUDIT IS NOT VALID, THEN IT IS ABSOLUTELY A WASTE OF TIME AND MONEY.

Is the design and conduct of a reliable and valid audit worth the effort and resources required? This is, of course, a business decision. However, in order to make an informed decision, the manager would need to begin with an understanding of the reliability and validity of the particular audit. Based on our reading of the literature, simply assuming that an audit is reliable and valid is a very unsafe bet. In addition to this knowledge, it would be useful to have some way to estimate the losses that might be prevented. Ultimately this is a matter of conjecture. However, we can usually make some judgment about the stakes involved. We would be much more likely to construct reliable and valid audits on the safety of space shuttle flights or the operation of petroleum refineries than we would for the operation of a law office.

Nevertheless, it is critical to realize that if the objective of the audit is to provide a measure of performance or a prediction of future performance and the audit is not valid, then it is a waste of time and money.

SOME CONCLUSIONS ABOUT AUDITS

It appears that audits can be made reliable. To gain this reliability appears to require some training of the auditor and ensuring that the scoring criteria are operationally defined. If audits are to be used in a company, the audit proponents and practitioners should test reliability constantly and should not publish results until demonstrated reliability is achieved.

The validity should also be tested whenever possible. Do the plants with the best long-term safety records have the best audit scores? In this test it is essential that auditors not know the long-term record of plants they are auditing. Otherwise, the validity test is biased.

We are inclined to believe that audits have some value. Both of us regularly take our cars to the dealer to have them go through the checklist. However, the role of an audit is usually to maintain conformity to rules, regulations, and standards. This is not unimportant. However, it is unlikely that the data obtained from audits will serve in a planning process to enable the development of targeted improvement actions.

In order to serve as a key element in improvement planning, an audit process would need to have reliability, criterion-based validity, and construct-based validity. This is a hurdle that does not appear to have been cleared up to this point.

Dr. Deming pointed out that "you cannot inspect in quality." His point was that inspections are outcome or results measures, not process measures. Therefore if your aim is to improve the processes producing the outcome, you need to measure the controlling characteristics of the process not simply the outcome. In safety terms, this means inspecting and auditing for static outputs of a process will yield information on how well the process is producing an outcome. Without construct validation, it will not tell you what to do to improve the outcome or even why you are achieving the results you are getting.

KEY POINTS

- An audit is an examination to determine whether some set of standards is being met.
- When audits are used to measure performance, the persons being audited may bias the results by putting the best possible spin on their statements.
- Although audits can be made reliable, this is a difficult process. It requires comparing the scores of independent auditors.
- There is very little evidence of criterion-related validity of audits. The only substantial evidence we found was a demonstration that maintenance audits correlated with accident rates in a set of 25 manufacturing plants.
- Audits may be able to tell us how the system is performing but usually provide little insight about how the system functions or how to improve it.

ENDNOTE

1. They receive equal weight in scoring. However, if one of the questions gets the same score in all firms studied while the other varies from firm to firm, the variable question actually takes on more weight, because it contributes more to the variation in final scores.

8

The Safety Survey

In 1993 the authors of this book met for the first time. Pat Ragan was safety director of a large chemical company that had grown in the previous several years through a set of acquisitions. The company had over 6,000 employees in the United States, distributed across over 50 plants. Only two plants had over 500 employees. Most were in the range of 25–50.

This small plant size presented a problem for safety measurement in the company. In 1993 the company was well along with the implementation of a total quality management implementation. Based on the recent training in variation and the use of control charts, the safety managers had begun to use *U* charts when looking at incident-rate measures. The wide limits of variation encountered with small plant sizes convinced them that incident rates would not be particularly useful as a measure of performance in the company, at least for small sites and for short time periods.

The safety director convened a high-level committee of safety professionals and plant managers to address the issue. Essentially, this was a process-improvement team focused on the problem of company-wide safety measurement. Aware of Dr. Deming's admonition that system change requires new knowledge from the outside, the committee brought in consultants and industry experts for a series of seminars. Brooks Carder was one of the consultants and was asked to provide a seminar on the theory of safety measurement based on his knowledge of total quality management.

WHEN MEASURING A COMPLEX PROCESS, IT IS IMPORTANT TO USE MORE THAN ONE METHOD.

The seminar included explanations of the concepts of reliability and validity. One of Carder's assertions was that, when measuring a complex process, it is important to use more than one method. If several methods are in some agreement, you are on the right track. If they are not in agreement, you need to find out why. Either way, you learn important information.

A review of approaches to the measurement of safety included a discussion of what were called *perception surveys*. These are written surveys of employees that include a variety of questions on safety and the safety-management system. We no longer use the term *perception*. These are like other opinion surveys, asking about the attitudes and beliefs of the employees. The assumption is that the actions of employees will be guided in part by their attitudes and beliefs. The same assumption holds when pollsters survey likely voters to predict their voting behavior in an election, or when marketers study customer attitudes in order to design an effective advertising campaign. These are not called perception surveys. Carder recounted some positive experience with surveys on attitudes and beliefs about safety. He had found such surveys to be a useful measure, particularly if the intent was to use the measure as a basis for planning process improvement.

Because the survey appeared to be the most promising new measurement available, the team made a decision to proceed with a pilot study of its use. Carder was asked to serve as a technical consultant to help the team evaluate the reliability and validity of the survey in the new context.

The most prominent survey at the time was the Minnesota Safety Perception Survey. This survey had been developed in the 1980s by a team that included safety professionals from the Association of American Railroads and scientists from the Army's Aberdeen Proving Ground. Dan Petersen served as a consultant to the team, which was led by Chuck Bailey (Bailey and Petersen 1989; Bailey 1997). Bailey's team designed and validated the Minnesota survey, which had subsequently been administered to over 100,000 employees in a number of companies. Bailey's database included survey results from six chemical companies, giving us a good opportunity for benchmarking. Consequently, we chose that survey. Chuck Bailey made a presentation to our team to explain how the survey had been constructed and how it could be used as a tool for improvement.

The survey consisted of 74 yes/no questions written by the Minnesota team. These had been culled from a larger group of questions through validation studies by Bailey's group during the development of the survey. Valid questions were retained, invalid questions were dropped. As far as we could determine, however, no question-by-question validation studies had been carried out since the initial development of the survey in about 1980.

The improvement process developed by Bailey's group was an excellent example of the planning phase of the PDSA cycle. The survey was administered to all employees, including managers and executives. The findings were then fed back to the workforce. The analysis looked for gaps or weaknesses revealed by the data. This included finding questions where the scores were low compared to benchmarks and questions where there was a large difference of opinion between workers and managers. Both were taken as indications of conditions in need of correction. Teams of employees, mostly hourly workers, were then convened to discuss these gaps and help understand what was behind them. Based on this, action plans were developed and improvement processes were implemented. In 1993 there was nothing published in a peer-reviewed journal that demonstrated the effectiveness of Bailey's survey as an improvement method, but there was considerable anecdotal evidence that this process was effective at reducing incident rates.

To test the criterion-based validity of the survey, we designed a pilot study. The basis of the study was the assumption that, if a question was valid, then a site with an effective safety program should have a higher proportion of positive answers on that question than would a site that had a weak safety program.

Initially, we chose seven sites for pilot study. The sites were of relatively equal size and were judged to have relatively equal risk levels. Four of the sites chosen were judged to be excellent programs. To be in the excellent category, a site had to have a recordable rate (RAIR) that was below the division mean for the past three years and be judged by safety professionals to have an excellent program. (Two of the four excellent sites had received the President's Safety Award in the prior two years.)

Three of the sites were judged to be below average. To be in the below-average category, a site had to have an RAIR that was more than 2 standard deviations above the division mean in two of the past three years. In addition, we asked for confirmation from a company safety professional that the program was in need of improvement.

At each site a team was established to inform workers about the survey and see that it was administered properly. One important concern of our team was that worker confidentiality be protected. Therefore, surveys were not signed. They were usually filled out in groups at safety meetings and put into a ballot box. From there they went to the consultant. The consultant agreed not to provide breakdowns of data for any group smaller than five employees, so that particular employees could not be singled out in small departments.

Part of the context here was a pervasive negative attitude about surveys in the organization. In the prior year, the company had conducted a communications survey that was very long and tedious. It was then nine months before any feedback came from senior management about the results of the survey. In the view of the workforce, no action was ever taken on the findings. Therefore an important part of our commitment was to get data back to the sites and the employees within a few weeks. We could not promise that the company would act on the results, but our team understood that a failure to act would create a serious problem.

The survey we administered consisted of the 74 questions from the Minnesota survey, along with 14 additional questions written by our team. The additional questions were written to cover some issues not covered in the original survey but important to a chemical company, including process safety and emergency response. In addition we added some questions of particular interest to our team, covering incident reporting and the use of drugs and alcohol. Our plan was to conduct a pilot study and then roll out the survey to the entire company if the pilot study demonstrated validity. This was a classic application of the plan-do-study-act cycle. The planning included the team's educational process and deliberations. We were entering the study-act phase. The surveys were administered in the plants and returned to Bailey and Associates for processing. Within four weeks, results were returned to the team for our studies.

RELIABILITY

In our pilot studies, we tested only the validity of the survey, under the assumption that if it was valid, it would have to have at least moderate reliability. Subsequently, we quantified reliability in two ways. One method utilizes the split-half technique where a person's total response (percent of favorable answers) to one-half of the survey's items (randomly selected) is correlated with his or her total response to the other half. If a particular respondent has 67 percent favorable answers on one-half of the questions, we would expect a similar percent favorable from that person on the other half of the questions. Application of this method yielded reliability coefficients in the range of 0.9, indicating a high degree of reliability.

The second method for testing reliability was the test-retest method. This took advantage of the fact that we repeated the survey in most of the plants in several consecutive years.

Figure 8.1 is a scatter plot of scores for the same plants when tested in 1996 (POS96) and 1997 (POS97) using the same survey instrument. The correlation between the two test scores yielded a Pearson r of 0.82, indicating again a relatively high degree of reliability. If we were studying the reliability of an IQ test, we might want a somewhat higher number, under the assumption that IQ does not change materially over time. We would not want a perfect correlation with the safety survey. It is assumed that the

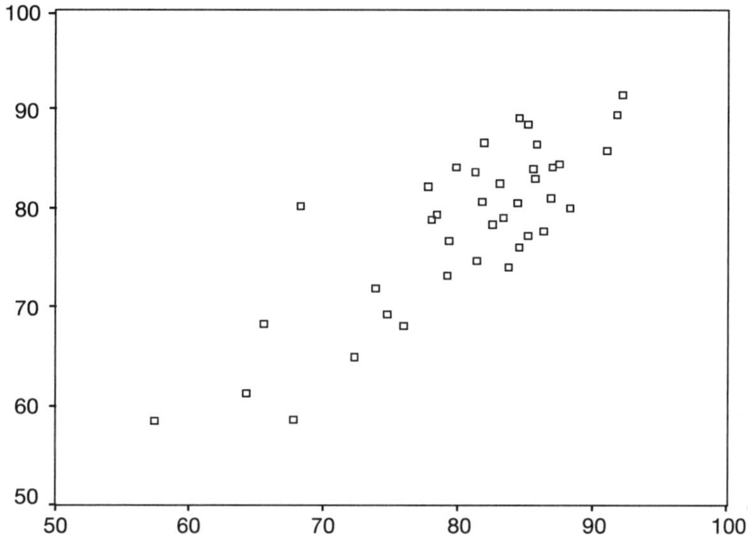

Figure 8.1 Scores of individual plants for 1996 and 1997 surveys.

performance of plants can either improve or decline, based at least in part on the actions of management. It should be expected that some sites will do more than others to implement changes based on the survey and that the scores of these sites will improve in comparison to the scores of sites that do not take significant action.

In addition, we inserted questions on subsequent administrations of the survey that asked whether the site has taken action on the previous year's survey. There is a positive correlation between the proportion of respondents saying that action was taken and the change in the survey score of the site. In one study of 23 sites for which we had data from 1996 and 1997, the Pearson r correlation between the questions on the site's response to the survey and the change in the site's overall score from 1996 to 1997 was 0.51 ($p < .012$).

VALIDITY

Content-Based Validity

The original Minnesota survey and the 14 questions we added were all written by teams of safety professionals. Because the judge of content-based validity would be safety professionals, the questions have *de facto* content-based validity. Some might argue that we are measuring only the perceptions of workers and that such perceptions may have no relationship to reality and, therefore, we should not assume content-based validity. Our response is that if workers have the perception that safety is not a high priority for management, then they are less likely to follow safety procedures. It does not matter what management actually thinks. Workers are guided by their own beliefs. They are unable to read the mind of the manager. Although there still may be a few extreme behaviorists who deny the influence of attitudes on behavior, the mainstream of psychology and economics would not dispute such influence.

Criterion-Based Validity

In spite of our arguments that incident rates are not an ultimate criterion, they offered an obvious and readily available criterion against which to assess the validity of the survey. In addition, we used the judgment of safety professionals regarding the quality of the safety program at the site. Here, of course, there were two levels: strong and weak. It must be realized that these criteria are not truly independent. When a safety professional is aware that a particular plant has a high incident rate, he or she would probably not judge it to have a strong safety program. A site with a low rate might be judged as weak, however, because the safety professional may be aware of serious hazards that the site has not effectively dealt with and conclude that they have been lucky so far.

The questions were initially validated using the following procedure: for each question there is a favorable answer and an unfavorable one. For example, for the question, "Is your training regularly updated?" the favorable answer is yes. For each question, responses are shown in 2 x 2 table format of Figure 8.2.

Each cell contains a count of answers that fit that category. If the question is valid, the excellent sites should have a higher proportion of favorable answers. This was tested using a Yates-corrected chi-square statistic

	Favorable answer	**Unfavorable answer**
Excellent sites		
Sites needing improvement		

Figure 8.2 2 x 2 table for question validation.

with a requirement that the one-tailed statistical significance be at or beyond the .05 level. We used a one-tailed test, because we were predicting the direction of the difference.

This study found that some of the original questions from the Minnesota survey were not valid in our test environment, even though they had been validated using a similar procedure when the Minnesota survey was developed (Bailey, personal communication). We subsequently conducted validation studies in another company. This second company manufactured copy machines. Looking at both studies, we found 10 questions from the original Minnesota survey that were not valid in either company. These are displayed below, numbered according to their position in the Minnesota survey.

6. Does the company actively encourage employees to work safely?
13. Have you been asked to perform unsafe operations?
16. Are employees provided information on accidents?
35. Has the Employee Assistance Program (EAP) helped eliminate alcohol and drug abuse?
40. Does drug or alcohol use increase incident rates?
50. Do safety meetings have an effect on safety performance?
51. Are employees with personal problems effectively handled by supervisors?
52. Do you have problems obtaining support for correction of hazardous conditions?
65. Are new employees assigned to experienced employees for instruction?
70. Is promotion dependent on good safety performance?

There can be a number of reasons why a question might be valid. First, with a larger sample or a different population, these questions might have proven valid. In our population, an obvious possibility is that the employees did not clearly understand the question. For example, based on our focus groups, question 13, "Have you been asked to perform unsafe operations?" seems to be interpreted very differently by different employees. If a question is not understood, or if it is understood differently by different workers, this would interfere with validation.

Another reason for failure to validate is that the scores are uniformly high. This is the case with questions 6, 40, and 65. Because even the weak sites score very high, there cannot be a statistical difference. Consider question 40. In 1980 some railroad workers might have believed that drug and alcohol use did not increase incident rates. In the 1990s it would have been difficult to find any group of industrial workers who were not nearly unanimous that drug and alcohol use increase incident rates. Over 90 percent of workers in all the plants surveyed answered yes to this question.

Finally, some questions received uniformly low scores. This is the case, for example, with question 35. Employees are not widely confident that the EAP is effective in eliminating alcohol and drug abuse, even in the best sites. Again, this could be because workers in the 1990s had more experience with, and knowledge of the effect of, EAP programs on alcohol and drug abuse than did the samples tested in 1980.

Although Bailey (personal communication) has reported that questions would be valid only when the scores of hourly workers were used for the validation test, our studies showed that most questions would also validate when only the scores of managers were used. Although managers generally had a higher proportion of favorable answers, these scores usually show the same pattern of strengths and weaknesses. We have subsequently observed that when there are severe problems at a site, managers' scores may actually be lower than those of hourly workers.

Having studied the individual questions, we carried the analysis further. We calculated a total percentage of positive responses by hourly employees to questions 1–74 for each of the seven pilot sites. These are displayed in Table 8.1.

Although the difference in survey scores between the sites with strong records and those with weak records is not large in terms of percentages, it is statistically significant, in spite of the small number of cases.[1] It should be noted that the statistical tests take the small sample size into account. In fact, to find significant differences with small samples requires a more reliable measuring device than if large samples are used.

Table 8.1 Percentage of positive responses to questions 1–74.

Site code	Safety record	Survey percent favorable
78	Weak	55
15	Weak	73
17	Weak	70
19	Strong	83
38	Strong	83
58	Strong	76
75	Strong	78
Entire company		68

Two of our weak sites scored slightly above the average for the company. (One reason for this is that the company had a very large site which had a low score, skewing the company average lower.) All of the strong sites were above the average, and there was no overlap between the groups. *The two highest positive scores came from the two sites that had won the company's President's Safety Award in the past two years.*

An alternative method of looking at the relationship between survey scores and incident rates is to compute a coefficient of correlation.[2] In order to do this, we added an additional six sites, three with good records and programs and three judged to have weak records and programs. Again, the sites were relatively equal in size and judged to have a relatively similar degree of risk.

Table 8.2 depicts the average recordable accident rate for 1992–1993, and the survey score (total percent favorable) for each site. In addition, there is a column entitled "rating." This is the rating given each of the sites by the company's North American Manufacturing Strategy Team (NAMST), based on their analysis of the quality of the management system in operation at the site. This was not an attempt to rate safety management but to rate the quality of the management system for manufacturing in general. The higher the score, the weaker the management system was judged to be. Because these scores were available, they offered a third criterion against which we might evaluate the survey. Although the accident record is not independent of the safety professional's rating of the site, this NAMST score should be independent of both.

Table 8.2 Average recordable accident rate for 1992–1993.

Site code	Perception survey	NAMST	Average RAIR
15	0.73	8	4.95
17	0.70	8	5.15
19	0.83	2	0.00
27	0.83	4	1.80
35	0.82	8	6.00
38	0.83	2	3.70
44	0.78	5	2.55
56	0.69	8	3.30
58	0.76	4	2.55
72	0.62	6	4.90
75	0.78	5	2.65
78	0.55	8	8.25
80	0.71	4	1.80

We then computed correlation coefficients to look at the relationships between the survey score, the RAIR rating, and the NAMST rating. Table 8.3 is a matrix of these correlation coefficients.

There is a negative correlation between the RAIR and the survey score, as we would expect. The more positive the survey score, the lower the RAIR should be. A correlation this strong would occur in random data only 19 times in 1,000, so this correlation is statistically significant and provides further evidence for criterion-based validation.

It is very interesting to note that the NAMST rating is also correlated with both the survey score and the RAIR. The correlation between the survey score and NAMST is negative, because a high survey score and a low NAMST score are both indicative of good performance. The correlation between NAMST and RAIR is positive, because the lower the NAMST and the lower the RAIR, the better the site. The difference in the strengths of these correlations may or may not be important. What is clearly important is that the existence of a correlation among all three measurements indicates that they must involve some common factors. The fact that the NAMST rating, which is not based on safety, correlates strongly with two measures of safety, suggests that we are dealing with some more general characteristics of the each plant.

ONE COULD ARGUE FROM THESE DATA THAT QUALITY AND SAFETY ARE FUNDAMENTALLY RELATED.

One could argue from these data that quality and safety are fundamentally related and that the survey, in its most general sense, is a measure of the effectiveness of the management system and is not isolated to safety. The NAMST rating was an attempt to measure the quality of the management system. Although it is not surprising that this is a strong predictor of safety, it is the only statistical evidence we are aware of that demonstrates the connection between safety and the quality of the overall management

Table 8.3 Correlation of coefficients for survey scores, RAIR ratings, and NAMST ratings.

	RAIR	Perception survey	NAMST
RAIR	1.000		
Perception survey	–.639 $p < .019$	1.000	
NAMST	.756 $P < .003$	–.577 $P < .039$	1.000

system. As a practical matter, the survey was a much less laborious process than the NAMST ratings, which took a team of two to four staff members several weeks at each site. The survey may be a much more cost-effective method to evaluate the quality of the management system with regard to quality and productivity as well as safety. In fact, we have sometimes edited these questions to remove the safety specificity and found them to be an effective tool in helping us understand the strengths and weaknesses of the overall management system.

To put the strength of these correlation coefficients into context, remember the example from another field, that the correlation between SAT and Yale grade point average is in the range of .2 to .3. This means that the SAT, the best available predictor, can only account for about 10 percent of the variation in college GPA. Compared to the SAT, the perception survey appears to be a much better predictor, accounting for 25–50 percent of the variation.

There is one problem that needs to be addressed here. Correlations do not specify the direction of cause and effect. The foregoing discussion has carried the implication that the attitudes and beliefs measured by the survey influence behavior in such a way as to make accidents less likely. In fact, the reverse might be the case: at a site with very few accidents, employees might begin to form more positive attitudes and beliefs about the safety system. Thus the results may be driving the attitudes, rather than the attitudes driving the results.

We have some evidence that bears on this question. We conducted a limited version of the survey at a pipeline company. Our finding was that management commitment to safety was perceived as weak, and recognition of employees for their contributions to safety, a component of management commitment, was lacking. Based on this finding, the company implemented a recognition program and increased management's visibility in supporting safety. Eight months later, we resurveyed their employees. The changes in survey scores are reflected in Figure 8.3.

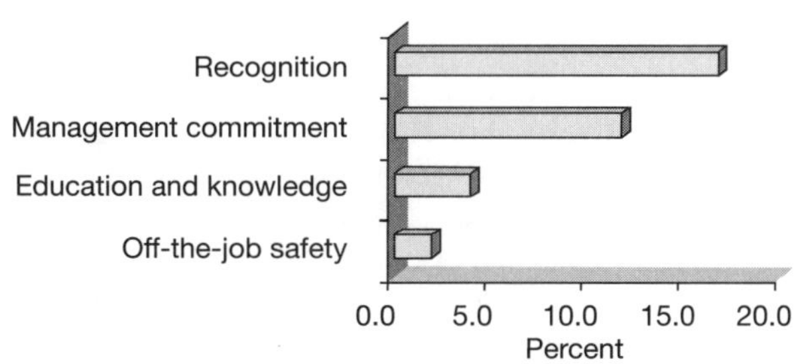

Figure 8.3 Improvement in survey scores.

Following the intervention, this company experienced a long-lasting reduction in the recordable accident rate in excess of 50 percent. We did find some general improvement in attitudes and beliefs. However, the largest improvements are specific to the areas targeted by our intervention. The improvements in recognition and management commitment are statistically significant; the other two changes are not. Moreover, the intervention was effective in improving performance. This is consistent with the notion that the survey score and the incident rate are both reflections of the safety-management process. We intervened in the management process and produced change in both the incident rate and the survey score.

The survey might be said to measure the *quality of the safety-management process* in a plant. This process influences the accident rate. However, when we think about the safety-management process, it is obvious that it is very complex. Managers and safety professionals could create a very long list of components of the process. The list might include: management commitment, employee involvement, audits and inspections, reward systems, training and education, emergency preparedness, medical facilities, safety communications, safety meetings, safety goals, and so on. We might presume that a valid test, like this survey, must be measuring some or all of these things. In fact, we could make our list of components and then go through the survey and decide which question goes with which component of the safety process.

The original Minnesota survey purported to measure 20 factors influencing safety performance, based on just such an analysis by the survey's authors. Initially we attempted to verify these 20 components by measuring the correlations among questions. More than one question is identified in each component. The logic of this is that if you have three questions that measure goals for safety, the answers to these three questions should have a relatively strong correlation. We found that questions within one of the 20 components often correlated more strongly with questions in other components than with other questions in their own component.

The 20 components were based on what the designers of the survey were asking about. We should realize, however, that the employee answering the questions does not necessarily have the same viewpoint about what is being asked. To illustrate, let us provide an example from some customer-survey work we did a number of years ago. The company's survey asked: (1) How would you rate our company overall? (2) Are you satisfied with our products? (3) Are you satisfied with our service? (4) Are our employees helpful? (5) Are our employees friendly? (6) Are our prices fair? The intention of the survey is to evaluate different components of the customer relationship. When we analyze the data, however, we frequently find that if you know how a customer answered the first question, you can predict very accurately how he or she answered the rest. Whereas you thought that you were asking about a number of components, customers often lump them all together in their mind and have some general concept of quality that they apply to your company.

In the safety survey, one of the questions is: are maintenance programs at a level that helps prevent accidents? Is this a question about maintenance, or about management commitment? Of course, the *content* is about maintenance, but what does the question mean to the employee? It turns out that employees view it as a question about management commitment.

Because we were not confident in the 20 factors proposed by the Minnesota team, we conducted a principal-components factor analysis of the database. We engaged a mathematical statistician, Marilyn Monda, to assist us with this process. A factor analysis is a statistical process that looks in the database for groups of questions that correlate. This process allows us to define sets of questions that are related to each other. We call these question sets *factors*. Our factor analysis indicated that the original Minnesota survey measured six factors. Rather than name the factors immediately, we decided to conduct focus-group research with survey respondents in order to better understand the factors.

Our statistical work indicated that the company was relatively weak in two factors: factor 1 that we tentatively labeled *safety leadership*, and factor 3, labeled *management involvement in the safety process.*

The weakness of the organization was reflected in two ways: Scores of hourly employees on the questions in these factors were low compared to scores of hourly employees in other chemical companies, and executives of the company had scores that differed greatly from the scores of hourly employees of the company. Executives in this case are senior managers.

We conducted focus groups at a series of four regional meetings: San Francisco, Houston, Chicago, and Cranbury, NJ. To each meeting, we invited four to five plant managers. Each manager was asked to bring one hourly employee. In virtually all cases, the plant manager himself attended, along with an employee. The format of the meeting was to begin with a presentation of the results of the survey that was similar to the presentation made to the company's executive committee. This described the history of the survey and our successful validation and outlined the weaknesses and strengths of the company, as revealed by the scores on various questions. Then the purpose of the meeting was explained. We then broke the participants into a group of hourly employees and a group of plant managers. Each group was asked to discuss the two factors, identify which questions were most significant, and decide what the factor meant. They were also asked to define the status of that factor at their sites and a goal of where the factor should be in the future, after a successful improvement effort.

The groups were brought together for feedback on their findings and discussion. They were again divided for a discussion of barriers to improvement. Again the groups came together, and we discussed the root causes of barriers, how these barriers relate to each other, and finally, possible actions.

WHAT THE EMPLOYEES TOLD US

The data compiled here represent the opinions of both plant managers and employees. Although both groups felt that the initial separation of the groups was helpful, they discovered that both groups had essentially the same opinions, although sometimes the perspective was slightly different. In no case did we discover an issue on which the two groups were divided. For example, if anything, the hourly employees tended to be more critical of hourly employees than were the managers.

Each group was asked to choose the *most important questions* for each of the factors. This was used as a method to initiate discussion and to attempt to focus the groups on the essence of the factors. The ensuing discussion was lively and open. The same questions were favored in each of the four regional sessions, so the data presented here represents the total response from the four sessions.

The questions from the Minnesota survey that were identified as factor 1 questions in 1994 were:

20. Does management insist on proper medical attention for injured employees?
25. Do supervisors pay adequate attention to safety?
26. Is safe work behavior recognized by supervisors?
36. Do employees participate in setting goals for safety?
37. Does company seek prompt correction of hazards found in inspections?
47. Is safe work behavior recognized by your company?
62. Have your company's efforts encouraged you to work more safely?
72. Are maintenance programs at a level which help prevent accidents?

The group selected questions 36, 37, and 47 as most important. Participants told us they chose these questions for two reasons: the question represented an important dimension of the safety system, and it identified a dimension in the company that was in need of improvement. Our statistical analysis of the survey demonstrated the greatest weakness on virtually the same questions.

Figure 8.4 shows how the company compared to six other chemical companies on the factor 1 questions. The negative numbers indicate that the company had scored below the mean of the other six companies on all the questions. The three most important questions are indicated by the large arrows on the left. They represent three of the four largest differences.

We also looked at the gaps between managers and employees, depicted in Figure 8.5. Here the larger the positive number, the larger the gap. Again, the most important questions are designated by the arrows, and again they represent three of the largest gaps. Thus, all three sources of data indicate the importance of questions 36, 37, and 47 as representing areas in need of improvement.

We carried out a similar analysis for factor 3. The questions from the original Minnesota survey that were assigned to factor 3 were:

33. Are minutes of safety meetings kept and follow-up files maintained?
45. Is discipline usually assessed when rules are broken?
53. Are checks made to be sure required protective equipment is used?
60. Do supervisors discuss safety goals and performance with employees regularly?
66. Are employees checked routinely to see that they are doing jobs safely?
74. Does company have uniform procedures for rules violations?

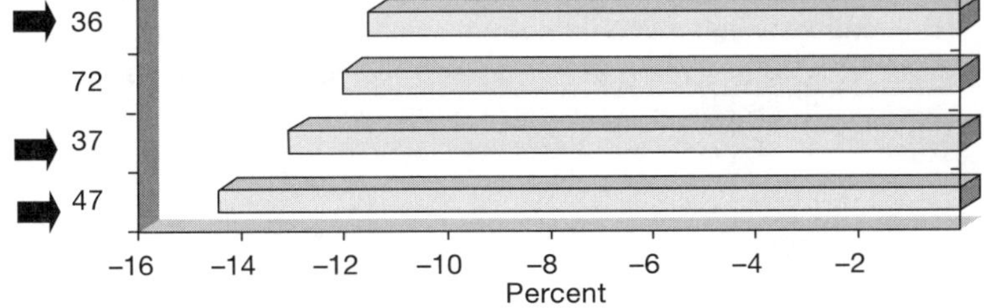

Figure 8.4 Difference from other chemical companies on scores of factor 1 questions.

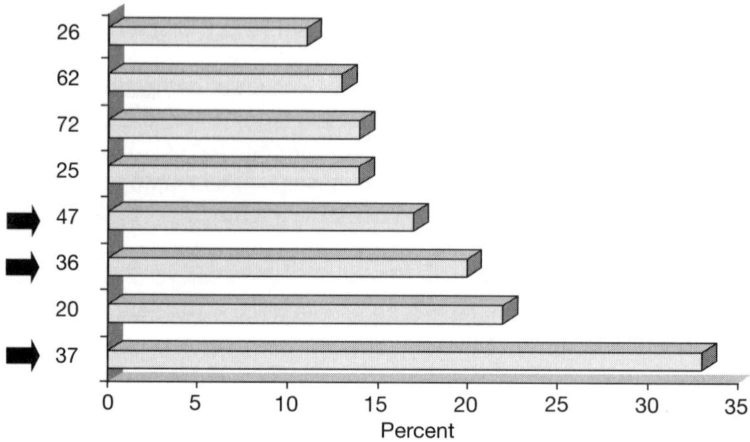

Figure 8.5 Gaps between managers' scores and hourly worker's scores on factor 1 questions.

Our participants selected questions 60, 66, and 74 as most important.

Again, we checked to see which questions on factor 3 represented the greatest weakness when the company was compared to other chemical companies. Figure 8.6 shows the differences between this company and the other chemical companies on the questions of factor 3, with the three most important questions designated by arrows.

Again, the larger the negative score, the larger the difference. Here there is a bit less correspondence between weakness in the scores and the employees' view of what is most important. Obviously, 74 is the weakest. Question 33 (Are minutes of safety meetings kept and follow-up files maintained?) is very weak, but it is easy to see why employees might think this is less important. On the other questions, the difference between this company and the other chemical companies is very small.

We also looked at the gaps between manager scores and employee scores on the questions of factor 3, depicted in Figure 8.7. In this figure we see large gaps between manager opinions and worker opinions on all three questions selected as important by our groups. Thus we have further confirmation of the importance of these questions to the assessment of the company.

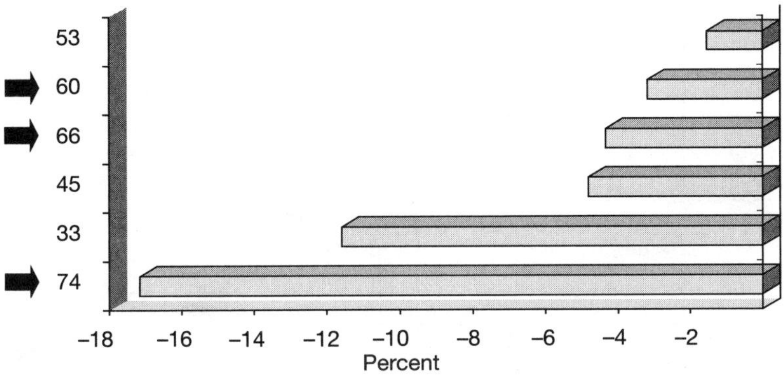

Figure 8.6 Difference from other chemical companies on scores of factor 3 questions.

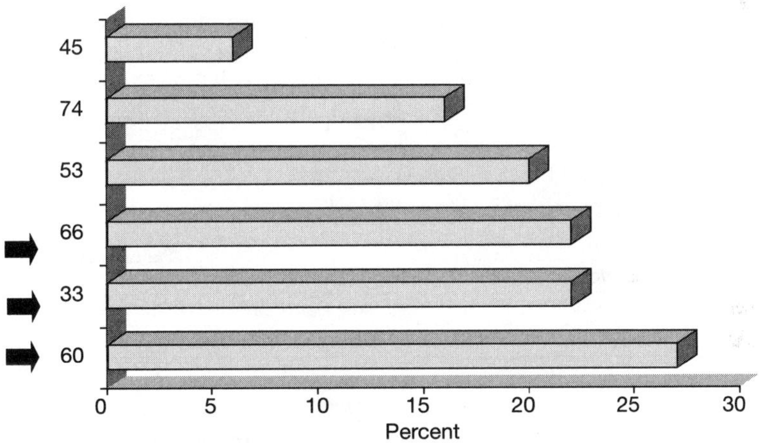

Figure 8.7 Gaps between managers' scores and hourly workers' scores on factor 3 questions.

NAMING THE FACTORS

Each group was asked to define what each of the factors meant. This generated much discussion and many attempts. Their suggested titles are summarized in Table 8.4.

Table 8.4 Suggested titles for factors.

Factor 1	Factor 3
Proactive accident prevention	No compromise–zero tolerance
Leadership	Is safety a concept or actual practices
Management of the safety process	Are the rules being followed?
Safety process implementation	Equal safety checks
Accident prevention	Standards
Management's commitment to safety	Work, practice, evaluation, and improvement
Employee participation	Ownership of the standard
Management's willingness to involve employees	Improvement
Management's demonstrated concern	Standards
Management's concern over a safe operation	Evaluating and maintaining safety standards
Is management doing its job properly?	Work performance evaluation and coaching
Management attitudes towards safety	Are we doing what we say we are?
Management commitment to safety	Maintain safety standards
Missing communication: managers–operators	Communication of commitment
Involvement from managers and operators as a team	Safety communication
Production vs. safety	Accountability and responsibility
	Rules and accountability
	Supervisors' interaction with employees
	The employee has to do
	Something has to control the process
	Procedures are followed?
	Measurement of safety performance and feedback

This is a difficult process. There is a tendency to jump to clichés rather than penetrate the issue. Many of the hourly workers came up with expressions that were verbally clumsy but were nevertheless the most insightful.

Based on examination of the records and our experience in the sessions, we decided on the following names for the factors:

Factor 1: Management's Demonstration of Commitment to Safety

Factor 3: The Quality of the Safety Supervisory Process

Some emphasis should be added here. Factor 1 is not management's commitment to the safety process, it is management's *demonstration* of commitment. Factor 3 is not the quality of the supervisor, it is the quality of the *supervisory process*. A supervisory process can exist with or without the existence of first-line supervisors. Self-directed work teams provide an example of a supervisory process that does not use first-line supervisors.

We convened similar focus groups and used the same methods to discuss the other factors. Based on these discussions, we arrived at the following six factors:

- Management's *demonstration* of commitment to safety. Do management's *actions* convey the message that safety is very important.
- Education and knowledge of the workforce. Are workers properly trained to do their jobs, and do they receive proper safety training? Do they understand their jobs and how to work safely?
- Quality of the safety supervisory process. Does the company have standards for work, and are these standards enforced?
- Employee involvement and *commitment*. Are employees involved in the planning process, and are they sufficiently committed to caution coworkers about unsafe practices?
- Drugs and alcohol (fitness for duty). Is drug and alcohol use prevalent and tolerated?
- Off-the-job safety. Does the company have an effective off-the-job safety program?[3]

The additional factors created by our new questions include emergency preparedness, process safety, and environmental protection.

In addition we wrote a series of questions specific to the safety philosophy espoused by the leadership of the company we were studying. These questions proved to be valid on testing. Because they were company-specific, we have omitted them from this discussion. However, they are of some interest in demonstrating potential uses of the survey tool. The fact that these questions were valid suggests that the philosophy was effective as a basis for the safety program. Had the questions not been valid, it would suggest that the philosophy was not effective. As a general practice, we allow each site to suggest a few questions to test the impact and effectiveness of site initiatives and programs. This has proven to be a very effective tool for evaluating these initiatives.

The first four factors appear to be relatively universal, being similar to the results of other attempts to identify the critical components of the management system. Table 8.5 depicts our factors along with the factors derived from three other sources: (1) our own factor analysis of a database from the application of a safety perception survey entitled the "Safety Barometer" developed by the National Safety Council; (2) a factor analysis conducted by Coyle et al. (1995) on data from a safety survey that he developed; and (3) the factors identified by a group of managers at Dow Chemical in the 1980s that formed the basis for the company's self assessment process.

The first four factors are present in all of the surveys except for the NSC safety barometer, where we found no component for education and knowledge.

In a similar vein, OSHA, through their VPP Star program, has identified four major elements of an effective safety program (OSHA 1989): management commitment and employee involvement; worksite analysis; hazard prevention and control; and safety and health training. Again, these overlap with our factors. According to OSHA, hazard prevention and control must include a clearly communicated disciplinary system. This is similar to our factor, quality of the supervisory process. Taking this into account, the OSHA elements cover all of ours.

Table 8.5 Management system factors from four sources.

Our survey	Safety barometer	Coyle survey	Dow self-assessment
Management's demonstration of commitment	Management's demonstration of commitment	Mantenance and management issues	Line management's leadership
Education knowledge	Not present	Training and management attitudes	Training
Quality of supervisory process	Quality of supervisory process	Accountability	Operating discipline
Employee involvement and commitment	Employee involvement and commitment	Personal authority	Total employee involvement
Off-the-job safety	Not present	Not present	Off-the-job safety
Emergency preparedness	Not present	Included in company policy	Not present
Drugs and alcohol	Not present	Not present	Not present

ESTABLISHING THE SURVEY AS A REGULAR MEASURE

With the adoption of the survey by our safety-measures improvement team, the intention was to use the survey annually or biannually. We were cautioned by experts about using the survey this frequently. They felt that there would not be sufficient time for change to take place, and there might be a practice effect in taking the survey. We felt this was an empirical question.

We administered the survey in 1994, 1995, 1996, and 1997 to essentially the same population. We did not observe a practice effect. Changes in most sites were indeed small, but they were real. When large changes were observed from year to year, we could trace the changes to some significant event, often a change in plant managers. The fact that the survey can be used on a regular basis is very important. Rather than serving as a basis for a one-time improvement effort, the survey can be incorporated into an overall measurement strategy.

EXPANDING THE SCOPE OF THE SURVEY

Because we were working in a chemical company, we had written questions on process safety[4] and emergency response. Of the 14 questions we wrote for the original survey, 12 proved to be valid. In 1995 we set out to further increase the scope of the survey. Using a team of experts from the company, we wrote questions on process safety and environmental protection and questions designed to measure the understanding of the company's stated philosophy and its adherence to that philosophy. In addition, we created three questions that asked about the site's response to the previous survey. (Of course, these questions could only be used at sites that had taken the survey before.) All of the questions were subjected to a validation study, using strong and weak plants, as judged through incident rates and ratings by safety professionals. To validate the environmental questions, we used data on releases and ratings by the corporate director of environmental protection to select a set of strong and weak plants. Each question was validated using the chi-square test on the 2 x 2 table in Figure 8.2. Most of the questions we designed proved to be valid. Our current survey has 96 questions, including 41 from the original Minnesota survey and 55 that our team developed and validated. Chapter 9 describes our experience with using the survey in planning an improvement process.

KEY POINTS

- The survey was adopted as a method to measure safety performance of small plants and to provide information for planning of process improvement.
- The survey was developed from the Minnesota Safety Perception Survey.
- The survey is reliable, with a split-half reliability of 0.90, which is quite high.
- The survey is valid when using criteria of incident rates and professional evaluations of the management system.
- Improvement actions targeted at weaknesses revealed by the survey appear to improve safety performance.
- Factor analysis and extensive focus-group work indicates that the survey measures fundamental components of the management system, including: management's demonstration of commitment to safety; education and knowledge of the workforce; quality of the safety supervisory process; and employee involvement and commitment.

ENDNOTES

1. The difference between the groups was evaluated using a *t*-test for independent groups. $T = 2.739$, $p = .02$. This means that a difference this large would be observed only 2 times in 100 if the two groups were not different.
2. A coefficient of correlation is an estimate of how well a score on one variable predicts a score on another variable. A coefficient on 1.0 means that the prediction is perfect. A coefficient of .7 means that knowing the score on one variable allows you to account for 49 percent of the variation in the other variable. (This is the square of the coefficient.) The associated probability score tells us how likely it would be to observe a correlation this high due to random variation, if the variables were not related. All three of the coefficients surpass the "5 times in 100 cases" criterion of statistical significance.
3. Our focus-group research found that employees consider this to be very important. It conveys the message that safety is not just an economic issue for management.
4. Process safety is essentially the management of hazardous chemicals. These are chemicals that are toxic, flammable, or both. Improper management of these chemicals can result in catastrophic results, such as occurred at Bhopal.

9

Taking Action Based on the Survey

The safety survey serves two objectives. First, it is a measure of the safety-management system and an indicator of future results to be expected from that system if the system remains unchanged. Second, the survey results provide a basis for developing actions for meaningful and efficient change. They provide information to assess the particular strengths and weaknesses of the safety-management system. When followed up with the appropriate improvement process, the survey has been shown to lead to significant improvements in safety performance (O'Toole 2002; Carder and Ragan 2003). This chapter describes the improvement process that we have used successfully.

ACTION AT THE CORPORATE LEVEL

In a corporation with several organizational units and many plants, it is best if each level of the organization develops action steps based on the findings of the survey for that level. Our safety-measurement improvement team was organized at the corporate level. After the survey had been validated and administered to all the sites, our team continued to function as an improvement team to develop action steps at the corporate level.

In Chapter 8 we described activities of a set of focus groups that helped us understand and name the factors. These focus groups actually had two tasks. After the naming of the factors, their second task was to look at the weaknesses that were revealed by the survey and help us develop an improvement plan. The process was a basic interactive planning process (Ackoff 1981), in which the group looked at the current condition, the desired condition, the gaps between the two, the barriers to closing the gaps, and the sources of those barriers. Based on the information developed by this process, the group then crafted strategies to remove the barriers.

The fundamental gaps identified at the corporate level were a weakness in *management's demonstration of commitment to safety* and a weakness in the *quality of the safety supervisory process*. The focus groups had discussed these gaps in detail while they were defining these two factors. Then we focused on barriers to closing the gaps. We facilitated the sessions. All participants were encouraged to give input, and almost all of them did. The challenge was for each team member to understand the input of the others and to build on that input. The objective was to create a dialogue in which meaning and ideas were shared, rather than a debate to establish who was correct.

The decision to initially separate the groups into managers and hourly workers had a very surprising and positive effect on the group. In the initial phase, the participants reported that the subgroup members were more forthcoming than they would have been in a mixed group. Managers would have inhibited nonmanagers and vice versa. When the groups were later combined, they discovered a remarkable similarity in their observations, much to the surprise of both groups. This revelation of how similarly their teams had understood the data and issues established a tone of trust and openness that greatly facilitated the later discussions.

BARRIERS TO IMPROVEMENT

The focus groups identified the following barriers.

Time and priorities. There was insufficient time or too many other things to do to accomplish safety improvement. Competing priorities included a focus on short-term financial results and production.

Cultural misfits. There was discussion in several of the sessions, especially from nonmanagement groups, about the barrier created by the company in allowing people to remain in the organization in spite of their continuing refusal to abide by the values and practices of the organization. At the simplest level, this was a failure to get rid of employees who willfully and repeatedly violated safety rules. The group felt that the continuing presence of these individuals created cynicism and discouraged other employees from engaging in improvement efforts. Many felt that the continued acceptance of this substandard safety performance was a symbol that the company was not really committed to safety and did not have clear, uniform standards.

Class structure. There was a perception of a class structure in which one class (hourly workers) did not engage in dialogue with another class (managers). This was defined as a barrier, but interestingly enough, given a proper setting, we had little difficulty fostering a dialogue between plant managers and hourly workers. However, there was wide agreement that there was a pervasive class structure in the organization, limiting willingness to communicate and cooperate on issues of all kinds, including safety.

The reward process. A number of observations were made about deficiencies in the reward process. The reward process focused on the short-term. It rewarded numbers not excellence, and it seemed to reward production over safety.

Chaos. This referred to the turmoil in the organization caused by such factors as the competitive marketplace and frequent reorganizations of the business. In addition, the group cited numerous instances in which their time was used to perform tasks that benefited corporate staffs but had no apparent benefit to their own operation. In our experience, some chaos is a basic property of business. This chaos is the business equivalent of the fog of war, and it is rarely absent. It can be an excuse for failure or a motivation to innovate and succeed. On the other hand, much of the chaos reported by our groups was unnecessary.

Lack of training and education. This primarily referred to training managers and supervisors in management skills, especially communications skills. It also referred to training nonmanagers in many of these same skill sets.

Empowerment without authority. Many instances were cited in which teams were given a problem to solve but no authority to implement a solution if they found one.

Supervisory process undefined. One of the reasons given for weakness in the supervisory process was that there seemed to be no definition or model in the organization. Because the company was built through a series of acquisitions, there was no unified culture, nor was there a companywide set of standards.

Fear. Fear, including job insecurity, fear of blame, and fear of punishment was seen to be a barrier to learning and change. Fear was frequently described in the sessions. It is important to make a distinction here. There can be fear of a person or persons, and there can be fear of the economic uncertainties confronting the company. In this case the employees were concerned that the company's economic performance might be weak, leading to plant closures and layoffs. These fears were realistic, as the company had already engaged in some layoffs.

No standards. Participants felt there were no clearly defined organizational standards for safety. More than one person pointed out that the supervisor is not the company, meaning that the message sent by the first-line supervisor may differ widely from the message sent by management. Also, different managers send different messages. One participant quoted an executive as saying that the safety effort should be "just the minimum that will keep me out of jail." This was certainly not consistent with the attitudes of senior management but illustrates the wide variation of standards in the company and why this issue was pointed out as a barrier.

Growing cynicism. Cynicism in this case is a distrust of authority, including an assumption that all communications from management are essentially self-serving and should be viewed with suspicion. This attitude is very destructive, as it impairs all communication. Cynicism, once established, is very difficult to overcome, because its maintenance requires only an occasional finding of error in management's communication. It was the opinion of the group that cynicism was present and growing though not yet pervasive.

Lack of communication and feedback. Here communication refers to the combination of talking and listening, with emphasis on the latter. Nonmanagement employees and managers alike felt that their upward communications are not heeded. A common observation, coming from many plants, was that work orders to repair hazardous conditions would disappear into the system. Often, the work was done, but there was no feedback to the person who submitted the request. Even when workers followed up, they frequently found no trace of the request.

These barriers are obviously closely interrelated. They comprise a system. Our next step was to take the barriers defined by the participants in our sessions and examine the interrelationships. The purpose of this was to find which barriers are fundamental drivers of the system. In other words, which barriers may be the causes of the others. It is on these fundamental causative factors, or drivers, that improvement efforts should focus.

THE RELATIONS DIAGRAM

The relations diagram we constructed is depicted in Figure 9.1. The logic of the relations diagram is to consider each pair of barriers and decide which causes the other. For example:

- Defining standards would be necessary to improving the reward process. Without standards, rewards would be seen as being offered on the basis of favoritism or even chance.

- Defining the supervisory process would probably reduce empowerment without authority, because it would define both responsibility and authority to act. Mismatches should become apparent. A lack of clear organizational standards adds to fear, cynicism, and organizational chaos. Again, without standards, employees will view the actions of management as being based on favoritism or luck.

The relations are not simple and are not always clearcut. The construction of the diagram is a process of dialogue. This dialogue is as valuable as the completed diagram, as it helps build an understanding of how the system functions.

In the diagram, an arrow points from a barrier that influences to a barrier that is influenced. In some cases, an arrow goes both ways. For example, the supervisory process is difficult to define without standards. At the same time, standards are difficult to establish without a defined supervisory process. Sometimes there is little or no relationship. For example, we saw no obvious relationship between lack of training and education and allowing cultural misfits to remain in the organization.

It is important to realize that this diagram represents the opinion of a particular group in a particular company at a particular time. It should be viewed as part of a process not an end result. It should also not be generalized to other companies without replicating the process that we describe.

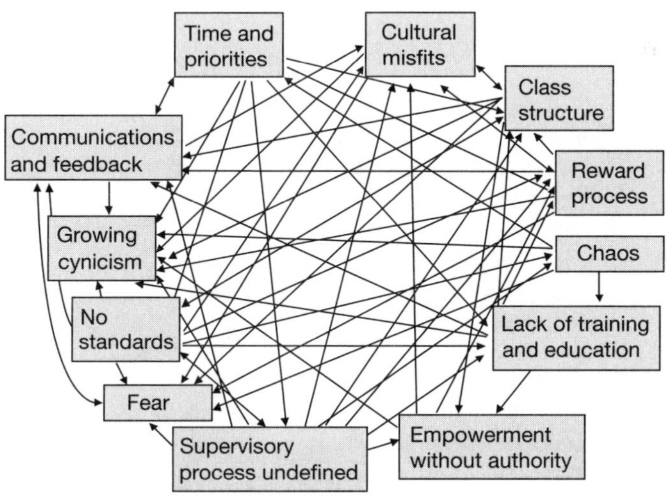

Figure 9.1 Relations diagram.

When the group's diagram was completed, we counted the number of arrows from each barrier and the number of arrows to each barrier. A barrier with a lot of arrows pointing out is a driver, influencing other barriers. A barrier with most of the arrows pointing in is a follower and should be positively influenced when we improve the state of the drivers in the system.

Figure 9.2 shows the net influence count for each of the 11 barriers. The bars going to the right, representing net arrows pointing out, are drivers. They are the initial links in the chain of causation. Changing them will change the system. The chief drivers are supervisory process undefined, no standards, and time and priorities. The bars with negative scores on the left side represent barriers with net arrows coming in and are followers.

The chief followers are fear and growing cynicism. It should be obvious that it is difficult to deal with fear and cynicism directly. It does not help to sit down with people and tell them not to be afraid or cynical. The only way to reduce fear and cynicism is to eliminate the conditions that cause them. For example, establishing clear standards for safety and enforcing the standards fairly would reduce cynicism about management's commitment to safety. Seeing management on the factory floor, listening carefully to what employees have to say, would reduce cynicism as well.

The reward process does not appear to be a powerful driver here, in spite of being the first place that many organizations focus on attempting to create positive change. The logical implication to be drawn here is that changing the reward process in the absence of defining standards and a supervisory process to maintain them would be ineffective. The implication

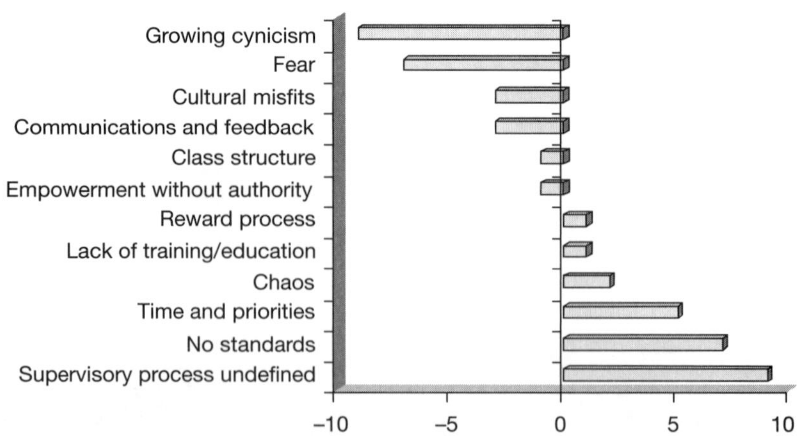

Figure 9.2 Influence count for barriers.

of the relations diagram, then, is to develop actions designed to establish standards and to define and implement a supervisory process.

EXECUTIVE ACTION STEPS

On the basis of the results described above, and using many suggestions derived from the group sessions, the safety-measures improvement team proposed the following five action steps at the senior executive level.

Establish Dialogue on Safety

A dialogue between levels of the organization should be developed. A dialogue is a two-way conversation, in which each party listens and attempts to understand the meaning the other is trying to convey. It is in the form of an informal conversation. Initial efforts would work best if supported by a facilitator experienced in this process. These conversations are initially difficult to develop in organizations that are relatively hierarchical.

The objectives of this dialogue would be as follows:

- A conversation in which executives actively listened to employees would be a demonstration of the priority of safety.
- The dialogue would be a forum for developing and communicating standards. Providing the employees with the opportunity to have input on these standards would strongly contribute to the commitment of all to implementing the standards. This would help develop constancy of purpose. A starting point would be a discussion of what we would like the standards to be and what they actually are in current practice.
- The dialogue provides executives with the opportunity to directly experience the attitudes and beliefs of the employees about the company. This information is essential to developing leadership strategy. It would go a long way toward eliminating the executive gaps we observed in analyzing the perception survey. (These were important areas in which the scores of executives differed widely from the scores of hourly workers.)

We recommended that each executive, from the leadership committee on down, participate in a two-hour facilitated dialogue at two plant sites during the year. In addition, we recommended that we share the results of the work of our safety management improvement team with the participants and at the annual health safety and environment (HSE) conference to simulate the dialogue and further demonstrate management's commitment to improvement.

Define a Safety Supervisory Process

We proposed to use the organization's established TQM process to define and implement a safety supervisory process. We would begin with an organizational unit that volunteered for the process. We would establish a team, including supervisors, hourly employees, and the plant manager. This team would develop, define, and document a process that could be rolled out across the organization.

The supervisory process had to have the following characteristics: it must be cost-effective and improve the overall efficiency of the organization; it should be based on the principles of TQM; and it must be flexible, so that sites with varying businesses and cultures can implement it.

Build Culture through Academy Approach

In order to establish an organizational culture in which standards and constancy of purpose would be strong, we proposed establishing an organizational academy.

In Chapter 10 we will discuss the transformation of the U.S. Army following Vietnam. A critical part of this transformation was establishing the training and doctrine command (TRADOC) (Kitfield 1995). This command was responsible for training and doctrine. Doctrine describes how the Army will operate and fight. In essence, it is a set of organizational standards. When the management of the Army adopts a doctrine, it is codified by TRADOC. Having written the doctrine, TRADOC also defines how this doctrine will be integrated into training all Army personnel, from the new recruits in boot camp to senior officers training at the Army War College.

Jack Welch used the academy approach at GE with great success, through the establishment of the GE training center at Crotonville. At Crotonville, Welch himself was a frequent participant in sessions with all levels of management. This was a powerful way to disseminate best practices and to support the establishment of a unified culture at GE.

Our academy would include curricula for all levels of supervision and management. It would provide managers with proper tools for management at each level. This would include communications skills (speaking, listening, and facilitating). It would include education in measurement methods, such as activity-based accounting. It would include education on systems theory and its applications in planning and problem solving. We proposed that this effort be consolidated with an ongoing effort to design and implement training for plant managers. We asserted that it was crucial that this be launched soon. Our goal was to complete the first training cycle for all levels of the organization in 18 months.

Stop Rewarding the Wrong Stuff

There was extensive discussion of the inadequacy of the organization's reward system to accommodate the improvements that are indicated. The reward system was too focused on the short-term and the superficial. For example, rewarding managers on the basis of accident numbers encourages nonreporting and ignores the long-term cultural issues that must be addressed for any long-term improvement. We proposed the following:

- Educate management about the effect of rewards on the entire system.
- Rethink the reward system. Our team would work with the human resources (HR) team to rewrite the organization's reward policy regarding safety. For example, we suggested using the survey to measure and reward cultural improvement. We believed, based on the results of our work, that survey scores were a more valid indicator of the effectiveness of safety management than were counts of incidents. This was true not only because of statistical issues (large variation in incident rates because of small organizational units) but because risk was not equally distributed across sites. We also recommended ending annual numerical safety goals for plants of fewer than 500 employees.
- Highlight improvement of the safety process rather than only improvements in the accident record in deciding on achievement awards.
- Understand how the organization's reward system perpetuates the class distinctions that are evident and alter them to promote dialogue between levels in the organization.
- Stop degrading safety with trivial incentives like safety bingo. Such games suggest that safety management is a game rather than a critical responsibility. We never see production bingo, or integrity bingo.
- Get rid of individuals at any level who repeatedly violate rules and interfere with the attempts of workers and managers to establish a positive safety culture. One of the clear demonstrations

to workers that safety is not a high priority is the fact that unsafe workers are not disciplined or fired. (Unsafe is defined as refusal to follow rules and procedures, not as having accidents.)

- Changing the reward system should be a priority for all management, top to bottom.

Repeat the Survey

Our work had established that the survey was a valid and reliable measure of the safety system. Given the number of small plants in the system, our team felt the survey would be a better measure than incident rates when considering periods as short as one year. In order to make the recommended changes in the reward system, the survey needed to be conducted annually. We felt the survey would be a leading measure, telling us how the safety-management system was likely to perform in the future unless we took action and helping us define actions for improvement. When looking at a potential investment, it is useful to see past performance, but it is more important to predict future performance. Should you buy, sell, or hold? The survey gave us a more reliable measurement by which to make these decisions about safety. For example, if we find a small plant with significant hazards and very low survey scores, we do not need to wait for a tragedy to know we need to take quick action to improve the safety-management system. The survey process will even tell us where to focus the improvement effort.

WHAT WAS ACTUALLY IMPLEMENTED

The first recommendation, to establish a dialogue on safety, was implemented effectively. Most of the senior managers in the company went to plants and conducted dialogue sessions that included hourly workers. These dialogues were recorded on video by professionals, and edited versions were shown widely throughout the company. In addition, Pat Ragan gave a report at the annual HSE meeting, which was attended by many hourly workers. This report was very similar to the report we had given to the executive committee. It clearly defined the weaknesses we had observed and outlined the teams' proposals for improvement. In addition to Pat's report to the whole conference of several hundred employees, the safety measures improvement team held smaller sessions, in which participants had an opportunity for dialogue with the team.

The second recommendation, to develop a safety supervisory process, was implemented but not to completion. One of the company vice presidents volunteered his business unit to serve as a pilot project. The project ran for about a year, with considerable progress. The vice president personally attended all of the sessions. However, corporate reorganization and mergers interfered before the process that was developed could be fully implemented.

The third recommendation, to build safety culture through an academy approach, was not implemented. A small subgroup had started work to put the academy together, and preliminary presentations had been made to management. But, because this was a longer-term project, it too was caught up in company mergers and reorganizations and did not get completed. A scaled-down version of this action was implemented at the largest site, and considerable improvement was seen. This step was a cornerstone of an improvement effort that has lead to five straight years of improving safety performance. In fact, it was not just safety performance that has improved but also production and maintenance performance results. We are not suggesting this was the only thing that caused these improvements, but it was generally agreed that this educational effort had considerable responsibility for the overall improvement.

The fourth recommendation, to stop rewarding the wrong stuff, was partially implemented. Survey scores became an important part of the performance evaluation of plant managers. Changes in survey results from year to year were considered to be a strong indication of the manager's effect on the site culture and on its future safety performance. Incident rates continued to be used but with appropriate statistical constraints.

The fifth recommendation, to repeat the survey, was implemented. In fact, in spite of mergers, acquisitions, and spin-offs of various entities, many of the plants that were in the original system continued to use the survey in 2001 and 2002. One of the major spun-off divisions later reinstalled the survey and the use of its results in its management objectives.

USE OF THE SURVEY AT THE PLANT LEVEL

We have described the use of the survey at the corporate level. The process of using the survey at the plant level is quite similar but typically demands fewer resources. In order to maximize the chances that sites would use the data effectively, we arranged regional sessions for plant managers to receive their data and participate in a seminar on what the data meant and how they could use it to develop a site improvement plan.

The process we recommended for developing targeted actions based on the survey data was as follows:

1. Survey results are fed back to the employees who took the survey.
2. Employee (management and nonmanagement) focus groups are convened to further understand the results and assist in developing focused action plans.
3. Plant management reviews the action plans. Agreement is reached on which action plans are to be implemented, and a schedule is developed.
4. Actions are implemented with clear support from plant management.
5. Results of the actions are measured and corrections made to the process as indicated.

PLANT STAFF VIEW THE SURVEY AS A USEFUL TOOL

Although we have evidence at the corporate level that the survey process is useful in improving safety performance, the opinions of operational management are also very important. If they viewed the survey and the related improvement process as useful, they were likely to make it happen. If they viewed it as not useful, they would not use it. In order to gauge acceptance of the process, we eventually surveyed a population of 22 plant managers who had used the survey annually for at least three years. Figure 9.3 shows their opinion on the usefulness of the survey for the company.

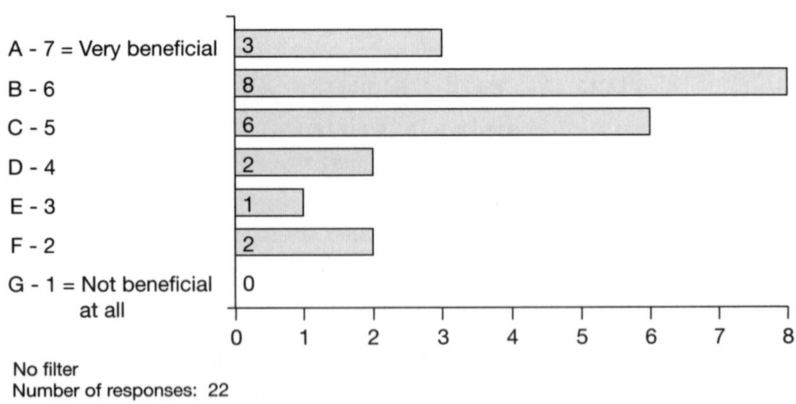

Figure 9.3 Plant staff opinions on the HSE survey's benefit.

The average score is 5.2 out of a possible 7. This indicates significant acceptance of the process. There is considerable variation here. Seventeen respondents found the survey useful or very useful, whereas five found it not very useful. Verbatim comments included the following:

> *[The survey] allowed the site to prioritize safety programs to meet the employees' concerns. For example in the first year we reemphasized emergency response, then off-site safety. In later years we have emphasized management of change and supervisory safety processes.*
>
> *[The survey] highlights latent problems of which we were unaware.*
>
> *[The survey offers] identification of areas needing improvement from the employee viewpoint.*
>
> *It provided a basis for developing the 1997 safety programs.*
>
> *Some good input was gained on several significant safety issues in the follow-up discussions.*
>
> *[The survey helps] to prioritize actions and empower groups to develop management systems in those areas.*
>
> *[The survey] builds safety awareness. Improves safety culture. Provides input from people other than mfg. Identifies problem areas. Can help to improve process safety.*
>
> *[The survey assists] buy-in to "safety" by nonmanagement employees.*
>
> *[The survey] helps get input so that HSE processes can be improved. Just conducting the survey and feeding back results communicates a level of commitment to HSE excellence.*
>
> *[The survey is] the best measurement tool currently available.*

In fact, in 1997 the survey was made optional, at the discretion of plant managers (though it was paid for by corporate). Over 90 percent of the sites elected to continue the survey.

EFFECT OF THE SURVEY ON COMPANY PERFORMANCE

Table 9.1 shows the change in the organizations' survey scores over the initial four years of survey administration. The five-point change represents substantial progress over the period, given the large size of the population under study. This could represent an improvement in the safety-management

Table 9.1 Organizations' survey scores.

Factors	1994	1995	1996	1997
1	72.3%	75.1%	75.5%	77.4%
2	76.5%	78.8%	80.2%	81.8%
3	62.5%	64.4%	64.4%	66.9%
4	86.4%	88.3%	87.7%	88.8%
5	65.9%	68.0%	64.2%	66.7%
6	79.0%	82.7%	80.2%	82.4%
7	63.0%	66.6%	68.7%	74.5%
Total (41 questions)	**73.8%**	**76.2%**	**76.6%**	**78.6%**

system or simply a practice effect, in which survey scores improve with repeated application, independent of any effective action.

Other evidence suggests that this represents an improvement in the safety system. First, not all sites improve. In Figure 9.4, each point represents the scores of a site on the 1996 survey (pos96) and the 1997 survey (pos97). This figure was presented in Chapter 8, without the diagonal line. Points below the line indicate that the site had a higher score in 1997, whereas points above the line indicate that their score declined from 1996 to 1997. Although most of the sites did improve, it is apparent that a number of sites actually did worse.

Moreover, whether a site's score improves or declines appears to depend on the actions taken at the site. In 1995 we inserted three questions to measure the site's response to the prior year's survey:

- Were you informed of the results of your site's 1994 safety perception survey?
- Did your site develop and implement actions based on the 1994 safety perception survey?
- Did your site benefit by participating in the 1994 safety perception survey?

We computed a composite score for these questions and looked at the correlation between a site's response to the survey and the improvement in the survey score. There was a significant positive correlation ($p < .03$) between response to the 1994 survey, as measured by these questions, and improvement in the site's survey score from 1994 to 1997. This, of course, does not prove that the survey process was causative. It may be that effective safety-management systems are likely to respond to the survey and cause improvement in survey scores over time.

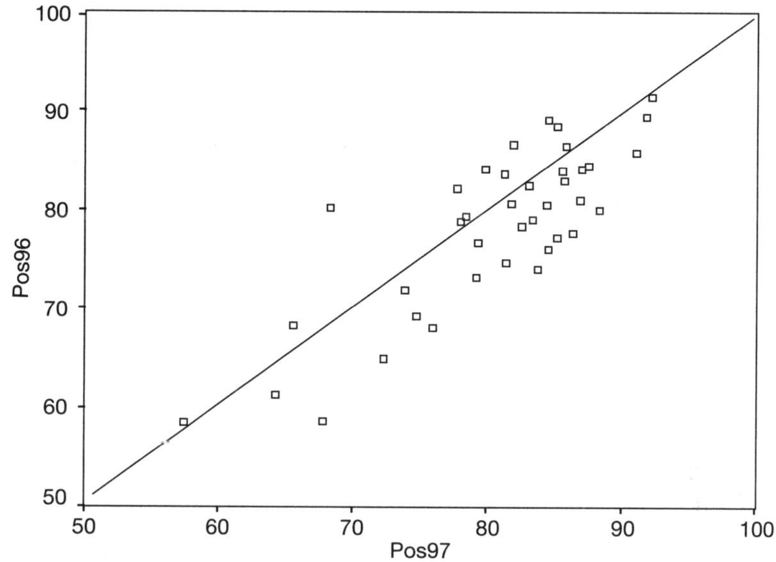

Figure 9.4 Scores of individual plants for 1996 and 1997 surveys.

The process appeared to have a powerful impact on the company's incident rate as well. Figure 9.5 is a control chart of incidents for the period prior to and following the initial application of the survey.

Five months after applying the survey, a process shift is indicated by seven consecutive points below the mean. The mean at the time of the survey was about 3.0, and after the shift, 2.2. This represents a reduction of over 25 percent. In our experience, this is a relatively small reduction. Figure 9.6 represents a single site of a smaller company that applied the survey and developed and implemented an action plan.

About four months after beginning the implementation of the action plan, they saw a shift in the mean recordable rate from about 10 to about 4.2. This represents a 68 percent reduction. Another investigator, using our survey and a similar process, saw a reduction in the lost-time rate of over 50 percent (O'Toole 2002).

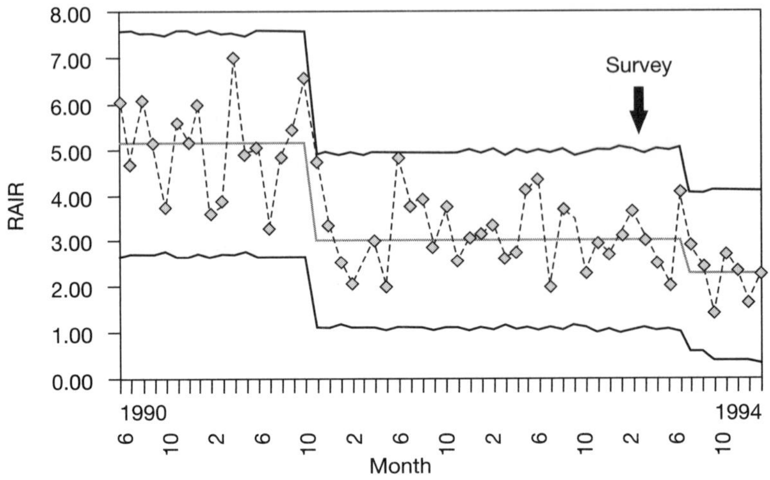

Figure 9.5 Incident control chart before and after applying the survey.

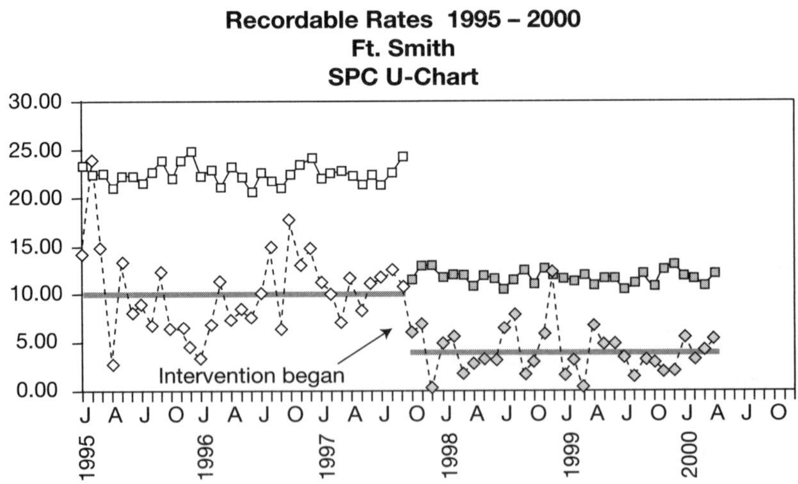

Figure 9.6 Incident control chart before and after intervention based on the survey.

CONCLUSIONS

The information generated by analyzing the survey provides a clear basis for an interactive planning process. We have described the version we promoted and used for that process, but many other formats would serve quite well. Our experience with the process defined here is that it is effective. We had success in changing the priorities of the company at the level of the executive committee, and the process was positively embraced by plant managers in the individual plant sites.

KEY POINTS

- The survey can be used to develop action plans at all levels of the organization.
- Focus groups of hourly employees and managers are convened to discuss the survey results and develop action plans.
- The focus groups discussed areas indicated by the survey as being in need of improvement. The discussion covers: gaps between what is desired and what exists; barriers to closing the gaps; relations between the barriers so that actions can be directed at the ones that are the most fundamental causative factors; and actions designed to remove the barriers.
- The process is effective at significantly reducing accidents.
- The process is embraced by both corporate executives and plant managers.

10

Leadership

Several years ago we received a call from the safety manager of a large manufacturing plant in the southern United States. He had become an enthusiast for the safety survey and wanted to use our survey at his plant. We conducted a survey of his 500 employees and analyzed the results. Later we traveled to the plant and presented the results to both managers and employees. When we departed, the safety manager was committed to following the prescribed process to develop and implement an action plan. We were a bit apprehensive about the situation. The safety manager was determined to act, and we felt he would push his position to the limit. We were not certain the plant manager had the same commitment, and we were worried about the friction that might develop.

About two years later we received a phone message from that safety manager. We actually thought that he was going to ask for help in finding a job. When we called him back, we were pleasantly surprised to find that our worries were for naught. His plant was one plant in a division of a much larger corporation. He told us he had been placed on a high-level corporate safety committee. "They don't much like what I have to say," he told me, "but they have to listen because our plant now has the best safety record in the company."

As we reflected on this chain of events, we began to realize the power of the survey in helping to make a case for change to senior management. We remembered the initial presentation that our safety measures improvement team made to the developing company's executive committee in 1995. We were expecting a difficult time then, too. We were bringing some bad news, and we were urging senior management to make some substantial changes in their approach to managing safety. We were particularly worried about the executive vice president. He had a reputation as being pretty tough, and we felt it was likely he would reject our arguments.

The two of us, along with the manager of one of the larger plants, made the presentation, which lasted about 45 minutes. We took them through our planning process, our validation of the survey, and our work with the focus groups. We explained what the work was telling us about the company, namely, that two major weaknesses were in the quality of the safety supervisory process and management's demonstration of commitment to safety. We then discussed what changes were indicated. Finally, we presented our list of recommended actions.

Before we finished our presentation, the CEO interrupted us and said, in a very friendly but serious way, "So what you're telling us, Pat, is that we're out of touch with people out there." Pat's answer was that no, *we* were not telling them that, the *data* were. We waited for a rebuff but none came. Finally one of us said, "Doesn't anyone want to argue?" I am sure whoever said this was looking at the executive vice president when he said it. The vice president simply asked, "How can you argue with data like that?"

In both of these examples, the information from the survey enabled individuals to lead with the confidence of a tested process and reliable data. It gave them credibility, and it gave them real insight into what was going on in the business. Although senior management could have rejected the information, it is our experience that such information generally is not rejected if properly presented. To be effective, this presentation must explain the validity of the survey and that the survey has enabled many companies to improve their safety-management process. The presentation of findings should be balanced to encompass both the weak points and the strong points of the management system.

Our original embracing of the survey grew from a concept of leadership that was derived in large part from Deming's thoughts on leadership. At the time we became acquainted with his work, Deming had over 60 years of life experience in observing and counseling leaders. In addition, Deming himself was a leader of considerable stature. Although no one reported to him but his long-time secretary, Deming led a transformation of Japanese industry that began in the 1950s and a transformation of American industry that began in the 1980s. It is hard to find a manufacturing plant of any significance anywhere in the world that does not apply many of Deming's principles.

In all five of his four-day seminars that we attended, Deming discussed the following attributes of a leader:[1]

Leaders understand how the work of their group fits into the aims of the company. The objective here is to avoid competition between departments.

Having an excellent safety record would not be useful if the company failed. Safety excellence should not be established at the expense of excellence in other areas. The previous chapters have provided evidence that safety excellence goes hand in hand with higher quality and lower cost. Each department and unit is there to further the aims of the company. Leaders need to communicate the appropriate role to their team.

The objectives of safety leaders are not best served if they are relegated to the role of the police or outside critics. They must be part of the organization's management team. This, of course, does not mean a compromise of important safety principles, but the realization that these principles must be part of the company's way of doing business. It must be built in, not tacked on. Incident prevention should be viewed by management as a key business advantage and an integral component of every process in the business.

Leaders focus on the customer, external and internal. I once had a very interesting experience at a conference of military safety managers. We raised the question of who their customer was. Very quickly the group mentioned the senior commanding officers of their respective services. Also prominently mentioned were the troops. As one of two nonmilitary personnel in a group of about 50, I waited with frustration and finally pointed out that the taxpayers should be considered a customer because we were paying for the services. This provided me with a visceral understanding of one of the fundamental problems of government: the customer who pays the bills, namely the taxpayer, is often left out of the equation.

In safety we have a number of important customers. The employees usually have the largest stake, because it is their lives at risk. The company's customers are also important customers of the safety process. They pay the bills. If safety is poorly managed, the cost of what they purchase will be elevated, and the quality of what they receive will be reduced. With many companies, the paying customer is also at risk. This is true in many service companies, such as airlines, cruise ships, hotels, amusement parks, theaters, and nightclubs. With companies that involve substantial risk of toxic release or explosion, the general public is a customer of the safety process. This is the case with chemical plants, refineries, and transportation companies that move hazardous products. Other safety customers include stockholders, management, and regulators. Deming said the first technology he taught the Japanese in 1950 was the customer survey. How often do safety professionals survey their customers to understand their wants, needs, and perceptions of what they are getting or how the incident-prevention systems are working? Deming also pointed out that although customer surveys are important, they do not tell us everything we need to know. Dr. Deming loved to point out that the customers cannot always tell you what they want.

- No customer asked for pneumatic tires. Charles Goodyear invented the vulcanized rubber (in 1844) that was later used for tires. In 1888, John Dunlop invented the air-filled or pneumatic tires. However, his tires were for bicycles. In 1895, André Michelin was the first person to use pneumatic tires on an automobile but not successfully. It took until 1911, when Philip Strauss invented the first successful pneumatic tire. Even then the customer had not asked for them. No one did a survey to ask, "Are the customers asking for pneumatic tires?" They just knew they had a better product and rolled it out, so to speak.
- In the late 1940s, surveys showed that many potential customers wanted to buy the Ford Edsel. When it was offered in the 1950s, most of them did not.
- Market surveys showed that the Sony Walkman would fail. It did not.
- One of the larger customer-survey mistakes in history was IBM's survey that told them not enough people would want computers in their homes to make it a profitable business. They did, and it was.

The point is not that surveys are not useful. They are. But there is much more to understanding the customer. Successful innovation is built on finding something that customers do not know they want but will desire when it is offered.

It is important to acknowledge that, historically, safety professionals have acted as if they had a captive customer base and could afford to take their customers for granted. Actually this is a bit of a stretch, because many safety professionals might have been hard pressed to say who their customers were if asked. An industrial company must have safety professionals to deal with the multitude of regulations and regulators. How well this was done seldom mattered, beyond the criterion of having better results than last year and no big fines by an agency. Safety departments were usually underfunded and underincluded in management processes. Today, that is under constant challenge. We are familiar with large companies that have put safety into a shared-services organization that is forced to sell its services to operational units. The units can buy the services in the marketplace and are not forced to use the internal services department. If the shared-services group does not provide the value the customer wants, the group will not maintain the revenue that supports it, and jobs will be lost.

In the instances we are aware of, this transition to self-supporting service organizations has not worked. One reason is that safety organizations could not make the transition from just telling their customers what to do and hoping they did it, to creating value that was evident to these customers. They also appear to have difficulty pricing their services to be competitive and financially sustainable. Another problem is that operational managers have difficulty changing this relationship. They are either overzealous in exercising their newfound ability to cut cost by cutting back on services, or they have difficulty managing the performance of these internal service groups. In our experience, managers usually felt they were being overcharged for the results they were getting, and perhaps they were.

Leaders are coaches and counsels, not judges. This is contrary to the traditional management formula that sets targets and holds subordinates accountable. Acting as a judge is an empty formula. What is added by the leader under this formula? Leaders acting as judges are limited by the extent of their own knowledge and ability. Teams lead by coaches and counselors can achieve more, because they rely on the abilities and ideas of the entire team to make progress.

The coach-and-counselor approach is especially important in safety, because safety professionals are usually expected to affect organizational processes and people who are outside their direct management control. It is the rare company that has a safety manager in charge of the manufacturing operations. But in almost all companies, the safety group is supposed to effect fundamental changes in these groups to lead them to having fewer incidents. This is very difficult when acting as a coach and counselor, but absolutely impossible if one occupies the adversarial stance of a judge. People are not inclined to submit to a judge who has no authority over them. However, they will begin to heed the advice of a coach who is able to demonstrate the utility of that advice.

One last note on the safety leader role as a coach and counselor. To be effective in this role, safety professionals must be perceived by their customers as professionals and experts. Safety professionals who are not seen in this light will have a very difficult time exercising effective leadership.

Leaders remove obstacles to joy in work. Deming argued that the most powerful motivation for workers is intrinsic motivation. Intrinsic motivation arises from the activity of doing something well, rather than the extrinsic rewards and incentives that the activity may generate. In spite of the fact that this is consistent with a great deal of research in psychology, many managers do not recognize the importance of intrinsic motivation in the workplace. I reflected on a job I had during the summer of 1960 at a Ford plant in Kansas City. The pay was very good, and no worker ever complained about it. Nevertheless, most hourly workers hated their jobs. This was not because of the monotony of the assembly line but because the workers believed the cars were very poorly made. My first job was to attach heater cables to the firewall under the dash. I received about 15 minutes of training. On many cars, I was able to get the screws in only part way, leading eventually to a customer who suddenly was unable to operate the heater.

Many of the workers talked about working at other plants where the quality was very good. They felt bad about the work they were doing and believed management did not care. We expect that management did care but are certain that management was out of touch and ineffective. Ford got pretty close to going bankrupt in the early 1980s before they brought in Deming, who helped management transform the company.

I had a different work experience that demonstrated the other side of joy in the workplace. During high school in Fulton, Missouri, I had a job at a tire shop. This was a difficult, dirty, seemingly joyless job. Depending on the season, the work was hot or cold. You can be sure we did not rush to work on many of the tires coming in from the local farm yards. But, and this is a very important but, the managers of this small business were involved with the work and the workers. They not only let us find joy in what we were doing, they joined in. When we were very busy, we turned the work into a game and had races to see who could change a truck tire the fastest or change and balance four car tires the fastest. Although we did not use the word "quality" at that time, quality was always important, too. If a tire went flat or the lug nuts were not tight enough or the wheel was not properly balanced, the one who did the work usually got to redo it. We did not like that. The owners also impressed upon us the importance of our work. If we did it wrong, we put our customers at risk.

I was mortified one morning when a customer I had serviced came walking back to the shop because one of the wheels I had installed had fallen off. It seems I had remembered the hubcap but forgotten the lug nuts. To this day I can remember dragging a jack up West Fourth, across Court and over to Business Rt. 54 with lug wrench in hand and lug nuts in my pocket. I had to jack the car up and put the wheel back on in the middle of my home town. Needless to say, several friends drove by and jeered. This is the type of experience that teaches you about the importance of a job well done and how important doing the job well is to safety.

When work slowed down and we got the shop cleaned up, we would stack tires. Even then we would work as a team and see how quickly and well we could do the job. When things were really caught up, which did not happen too often, we would have foot races or see who could throw a tire the furthest. Frequently, and especially after rush periods, the bosses would open the pop machine, and we would all have a soda together, sitting on the piles of used tires. I look back now and realize just how much that shop got out of a few free sodas, and I realize that it would have meant so much less if the bosses had not sat on those same tires with us when we drank them.

The impact was truly more related to the recognition associated with that occasion than to the cost or material value of a 25-cent bottle of pop.

This was a job that had more opportunity for obstacles to joy than most. By the way, I only made 97 cents an hour for this work. Somehow the two owners made a living for themselves and a few full-time employees, with enough left over to provide a job for at least one high school kid, who could earn a little cash for dates and walking-around money.

Later in my career I experienced the same dissatisfied, joyless work settings as the one discussed by my colleague at Ford, but this early experience of a closeknit work team, finding joy in their work regardless of what their work was, taught me an important lesson about the importance of joy in the workplace.

One might argue that it is easy to get this joy in a small tire shop but not in big companies. Every highly effective work group we have worked with since those high school days, in some of the largest companies in the world, has had some degree of this joy and excitement about what they were doing. The groups that did not have it may have been successful in achieving their task but never reached that same level of productivity or innovation. Even when we had the same group of people but a manager changed, or an acquisition happened, so that the barriers to joy crept back in, the group would lose some of the energy, some of the effectiveness, and some of the creativity we had before the change. We think the relationships built in the good times held and helped us be more productive than others who had not had these experiences, but with the barriers in place we were at 80 percent rather than 125 percent.

In today's corporation, there are many obstacles to what Deming called joy in the workplace. For example: (1) tasks are often broken up into such fragments that many workers never get the satisfaction of seeing their contribution to the finished product; (2) workers are often so hurried that it is very difficult to do a job as well as they would like; and (3) workers, particularly in customer service, are often so constrained by policies and procedures that they cannot do what makes sense to take care of the customer.

None of these conditions is necessary, and all create problems for the organization. Although specialization is necessary, leadership can and should provide everyone with a sense of the missions and accomplishments of the entire group. This should be done on an ongoing basis not just in annual review carried out at arms length by senior management. The accomplishments of a working group are occasions for celebration not for issuing a memo.

A great amount of organizational hurrying is the result of management's failure to clearly define priorities. This is not an argument that work should be carried out at a leisurely pace but rather that one task should be brought to completion before several others are initiated. Stephen Covey's (1994) *First Things First* addresses this issue with eloquence. He points out that there is a difference between urgency and importance, and that what is important is seldom urgent. To the extent that urgent things crowd out important things, our work and the work of our group will be less successful.

To provide an example of the urgent crowding out the important, consider a group of plant managers with whom we worked over a period of months in a planning process. The group acknowledged that the single most important management action was conducting informal discussions with hourly employees on the factory floor. This would give them information about how things were going that could not be found in written reports or in communications from direct subordinates. However, the managers spent little or no time doing this. It was crowded out by more urgent but less important functions like responding to information requests from various corporate staff units. If managers did not submit a report on time, there would be follow up. If they failed to talk to employees on the floor, no one would question it.

Leaders understand variation. The leader is in a position to make judgments, and, based on those judgments, to deploy human and financial resources. If data are available to guide a decision, interpreting the data requires understanding of variation. The manager who holds a special meeting when there is a one-month increase in accidents is wasting time and hindering progress, unless that increase represents an excursion beyond statistical control limits. Even then the meeting should be to understand what caused the accidents not affix blame for having them. A large proportion of the daily activities of most managers would be discontinued at once if those managers had a fundamental grasp of the laws of variation.

Early in my business career, I served for several months as a sales manager. Initially, I did what I was taught to do. Each morning I would review the sales figures. Some sales people would have had a good day or a good week, and I would call them with encouragement. Some would have done less well, and I would call them and very politely inquire about what might be wrong. Often the group with higher sales sold less, and the group with lower sales sold more.

Whatever happened, there was plenty of work for the sales manager. Of course, this work was counterproductive and a waste of time. The laws of variation make it clear that sales will vary from day to day. Treating this random variation as though I could determine the cause, without understanding the underlying processes driving this variation, will usually impair performance. I became more effective as a sales manager, meaning the sales team sold more, when I stopped bothering the sales force and began doing things that made selling easier for all of them.

When the management team of one of the companies we worked with understood variation of the accident processes, we discontinued the use of accident rates or counts for groups of less than 500 employees. Following this, the smaller sites were able to spend more time training people, improving procedures, and removing hazards, instead of responding to executive calls about why their rate had gone up or down in a particular month.

Leaders work to improve the system their people work in. There are two important points here. First, the focus of the improvement effort is on the system not the people, and second, the leader is a part of the system and not separate from the system. In the sales-manager story, working on the system could include getting better pricing, developing a stronger pitch, finding more attractive products, or developing an advertising program to support the sales effort. Sales training is another approach to system improvement, although a lot of sales training is a motivational program disguised as training. Motivation should flow from an effective sales process not from cheerleading by a motivational speaker. None of the actions our safety measures improvement team recommended to senior management was focused directly on the hourly workers. We wanted to change the system by changing the actions of management.

When our understanding of variation reduced tampering and gave us time to focus on improving our processes, we began to get results we never believed we could achieve. We got better and better at sending people home at the end of the day without injuries. Over a ten-year period, we reduced the injury rate by 80 percent. In the next seven years, we had reduced it by another 66 percent.

Leaders create trust. If trust cannot be established, the work of leaders is greatly impaired. Trust is evidenced by the willingness of people to undertake actions they cannot be certain will lead to the desired outcome, based on the advice of a person or group that is trusted. If the outcome is certain, trust is not required. Basically, then, trust is the condition under which learning and innovation can take place. Without trust, workers will attempt to stay with the tried and predictable. They will avoid innovation and risk.

Trust is also the glue that holds a group or team together. For a team to optimize its outcome, each member of the team is required to make some sacrifices. If team members try to optimize their personal outcome, the team will not play its best. In a basketball game, trust is required to pass the ball. The passer has to trust that when the circumstances are appropriate, the other player will pass it back.

Trust is easy to destroy, and after it has been destroyed, it is very difficult to reestablish. If people conclude that someone has deceived them, then all subsequent communications are viewed skeptically. Leaders in this position now have a nearly impossible task. Any error that they make in communication is likely to be viewed as another occasion in which they are attempting to deceive. Even very infrequent errors will be sufficient to maintain the lack of trust.

Trust is built slowly. As obvious and as simple as it might sound, a fundamental building block of trust is telling the truth. Just as important is the ability to listen and identify with the point of view of subordinates. This does not necessarily mean agreeing with that point of view, but it does mean that the views of subordinates can never be summarily dismissed.

Leaders forgive mistakes. Unforgiving leaders have some real problems. Because leaders are likely to make mistakes, and because those mistakes are likely to be more costly than the mistakes of subordinates, there is a great risk of unforgiving leaders alienating their subordinates.

Probably more important is the impact of fear on the behavior of subordinates. Relatively speaking, those who engage problems and make decisions are more likely to make mistakes than workers who duck responsibility. In organizational cultures that are intolerant of errors, we see an ever-increasing proportion of employees and managers who duck responsibility. This is a fundamental characteristic of bureaucracies that resist action. They are created when the penalty for error is greater than the incentive for successful innovation. In this circumstance, the low-responsibility path is the one taken.

In the example of the tire shop, when I returned to the shop from replacing the wheel on the car, not only did my boss forgive the mistake, he said he expected more mistakes, because he certainly expected me to continue to take on responsibility. What he said was, "The only people that don't make mistakes are those that never do anything." I could have been fired for this mistake. No one was hurt and the car was not damaged, but it could easily have been a serious accident. My boss understood that this did not represent a trend of absentminded mistakes. As a result of this experience, I was much less likely to make it again, compared to a new employee he might get to replace me. Tire repairman was not exactly a sought-after job. He acknowledged my mistake, let me know he understood that I had not meant to do it, and he reminded me in a sentence or two that it was still OK to try to do new things and improve things. More than 30 years later, I still remember and think about what I learned in that shop.

Based on what we know about psychology and on our experiences as managers, what Deming offers makes a great deal of sense. It is important to note that nothing is said here about incentives. Deming felt that incentives did not deserve the emphasis most managers place on them. In fact, incentives appear to have little to do with the performance of senior management. Most of us will recall that the senior management of Enron and WorldCom were among the highest paid executives in the world. The work of Jim Collins (2001), which we described briefly in Chapter 1, provides supporting evidence.

Collins studied the 11 out of 1,435 U.S. public companies that met his criterion for emergence into greatness. In understanding what enabled these companies to become great, it is perhaps most interesting to see what they did not have. Not one of the companies had a high profile CEO; they did not pay their CEOs more than competitors paid theirs, and they had no program to motivate the troops.

The motivation for the CEO and, presumably, for the troops came in large part from a vision and a set of principles from the hedgehog concept we described in Chapter 1. The concept is diagrammed on the left side of Figure 10.1. This is a concept for the leadership of an entire company. We believe this can be translated into the management of safety in a company. A safety equivalent is diagrammed on the right side of Figure 10.1.

Actually filling in the circles is not a simple task. It took the average company in Collins's study three years to work out their hedgehog concept. For example, Walgreens' decision to use profit per customer visit rather than profit per store as a measure of success was critical. To assume that the measure of success is something simple like a low rate of recordable incidents is a mistake. The task of our safety measures improvement team was to find better measures of success. The measures we selected were effective in helping improve the management system and prevent accidents in the company in which we worked. That is no guarantee that these particular

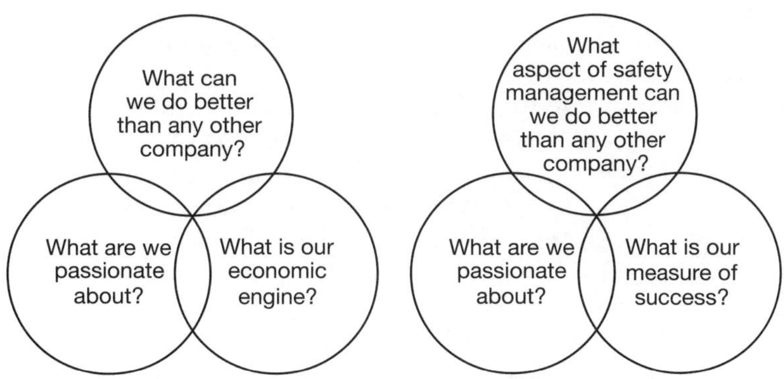

Figure 10.1 The hedgehog concept.

measures would be equally effective in other companies, although our experience with many sites in several companies suggest this is a good place to begin.

What we were lacking was the motivational element of having something in safety at which the company could excel. If anything, one of the companies had an inferiority complex. Our survey included some questions in which the workers compared their company with other chemical companies. Most thought their company was doing worse at safety than other chemical companies. We contrasted this with the survey results from a large plant at another chemical company. Although our company scored higher on the overall survey, it scored lower on the comparison questions. These workers underestimated the achievements of their company. The critical motivational element of the hedgehog concept, a realistic aspiration to excellence, was missing from our equation.

POWER AND LEADERSHIP

The ability to exert power and wield influence is not equivalent to good leadership. Hitler was a master politician and achieved great power. Stalin was less of a politician but was more successful than Hitler in gaining and holding power. In our opinion, neither could be considered a great leader. We believe, however, that to truly be considered a good leader requires an element of moral conduct. For example, Harry Truman, during his term of office, was quite unpopular. His power was diminished, because he faced a congress controlled by the Republicans for much of his tenure. However, he is now revered as a leader and is ranked second among modern presidents, behind Franklin Roosevelt and ahead of Ronald Reagan and Dwight Eisenhower (Lindgren 2000).

Many years ago I had the opportunity to meet Truman and experience him as a person firsthand. I had a summer job with a company in Kansas City that was owned by an old friend of Truman. When he learned that I was reading a history of World War II, he concluded that I "must meet Harry." He called and arranged the visit. On the big day, I had arrived a few minutes early. "The President is on his way," said the secretary apologetically. I was in the summer between my junior and senior years at Yale and was generally fairly impressed with myself. I was not really prepared for this, however.

About five minutes later, a short but very energetic man who appeared to be in his late 60s or early 70s (he was actually 79) strode briskly into the room. He stretched out a very firm hand with warmth and enthusiasm and apologized genuinely for his (two minutes of) tardiness. "Come on into my office," he said. The office of President Truman, at the Truman Library in Independence, Missouri, was large and magnificent. Books and papers were everywhere. Even at that age, President Truman was an avid reader and scholar. The office was full of memorabilia but appeared to be a place that he worked in not just a museum.

The president and I chatted for about 15 minutes. I explained that I was a premed student, entering my last year of college. "That's very good," he said. Then he said, "I never had a chance to go to college myself." He bowed his head slightly, looked very directly at me over his glasses, and said with a grin, "and I did pretty well." However, it was clear that he approved strongly of education. At the end of our chat, he showed me to the back door of his office, which led into the library and museum. To the left of the door, on a little shelf on the wall, was a guest register. "Please sign this," he said. Then he continued, "When Dwight Eisenhower came to see me, I asked him to sign this . . . in case anything turned up missing."

This man, reviled in his time as president, was a man of great character and strong moral compass and was a great leader. He was a man raised in a strong culture of segregation, who insisted on integrating the officer corps of the armed forces, against fierce opposition from the senior officers. When his business went bankrupt during the Depression, he worked for years to ensure that he paid back every creditor in full, even though the bankruptcy filing of the business had technically discharged his obligation. A Democrat whose constituency was organized labor, he nationalized the steel industry to prevent a strike he felt would be very damaging to the nation. He knew that what he did was likely to be overturned by the courts, but he also knew that by the time the judicial process was finished, the issues would be resolved and the strike prevented. As president, he attended the funeral of his old friend Tom Pendergast, against the advice of his top advisors and in spite of certain negative publicity. Pendergast was the fallen party boss who had reigned at the top of a very corrupt political machine in Kansas City. Pendergast had given Truman his start in politics, although Truman was never a part of Pendergast's organization and was never implicated in any of Pendergast's illegal activities. But he was Truman's friend. To skip the funeral would have been politically wise, but it would have betrayed a friendship.

Although our meeting was brief, Truman impressed me with his energy and humility. He listened carefully and patiently. He spoke in a very personal way, with warmth and humor. Interestingly, he did not project that personal warmth on television. I recall his speeches, which we can still view on videotapes, on the black and white television of the era. They were dry and cold. It is unlikely he could have been elected in this age of powerful mass media.

I also personally encountered Mr. Truman's generosity and concern for others. Following my visit, he wrote a letter to my employer, noting that we had had "a nice talk." Of course the president knew I would end up with the letter, which is now my most prized possession.

What made Truman a great leader? Obviously he was very bright and, though he did not go to college, a serious student of history. However, two things were much more important. As a humble man, he was able to continue learning. One of the critical qualities that Collins found in the CEOs who were able to transform companies was humility. These CEOs were determined but never arrogant.

The second quality that distinguished Truman from most politicians and most CEOs was that he was always willing to put his money where his mouth was. He made hard decisions and suffered the consequences. He left the glory to others while he took a great deal of blame. The sign on his desk said "the buck stops here," and it did. But he often let the credit for good deeds stop elsewhere. His plan to rebuild Europe was called the Marshall Plan, because he knew that Congress was much more likely to approve a plan named after a popular secretary of state than one named after an unpopular president. Truman's dedication was to the institution not to himself.

You are likely thinking that this may be a nice story, but what does it have to do with safety? We argue that it has everything to do with safety. Safety leaders who are willing to put their money where their mouth is and lead by personal demonstration, and who listen to workers and colleagues with humility and an ability to learn, will be effective. In fact, the findings of the safety survey strongly support this. When we first looked at the questions of factor 1, it was apparent that these questions were dealing with leadership. These questions provide, then, an empirically based way of defining leadership.

QUESTIONS THAT MEASURE LEADERSHIP

- Is safe work recognized by supervisors?
- Do employees participate in setting goals for safety?
- Do supervisors pay adequate attention to safety?
- Does the company seek prompt correction of hazards found in inspections?
- Are maintenance programs at a level that help prevent accidents?

- Do different departments work together to improve the safety of processes?
- Are your processes properly designed to operate safely?

These questions measure management's active and personal demonstration of support for safety. This includes management's allocation of time and money. It includes employee recognition. It includes involving employees in setting organizational goals, which of course requires listening to what they say. In each of these areas, management is demonstrating that it is part of the team. It is willing to makes its own sacrifices to meet the organization's goals. Our focus groups named this factor "Management's demonstration of commitment to safety." This is the essence of safety leadership. Harry Truman demonstrated his commitment to the United States of America by what he did not by what he said.

TO THE PEOPLE IN AN ORGANIZATION, MANAGERS' COMMITMENT TO SAFETY IS NOT SO MUCH ABOUT HOW MUCH THEY TALK ABOUT IT. IT IS MUCH MORE RELATED TO HOW THEY LISTEN AND ACT.

In many organizations, we encounter managers who wonder why their support for safety has not resulted in the desired improvement in safety performance. When we inquire into how these managers demonstrate their support, we usually find that they talk about safety, believe in safety, or are committed to safety. This is not enough. None of the questions that measure leadership asks if managers *talk* about safety. The questions do ask how much managers *listen* about safety. They also measure management's willingness to commit resources and to recognize contributions from employees.

LEADERSHIP IN THE ARMED FORCES

One of our favorite books on leadership is *Prodigal Soldiers* by James Kitfield (1995). The book chronicles the transformation of the U.S. Army from its defeat in Vietnam to its excellence in Desert Storm in 1991. During the late stages of the Vietnam War, the morale in the Army was very low. A group of faculty members from the Army War College conducted a formal study of this problem. The study can be found on the Internet under the title "Study on Military Professionalism" (Army War College 1970.) They interviewed 450 officers at the rank of captain, major, and lieutenant colonel. The researchers believed that this group would have a good picture of what was happening but would be too junior to be part of the problem. This is very much the equivalent of surveying hourly workers in a plant. They have not made the policies or the decisions, but they are aware of the results of the policies, certainly regarding safety.

Vietnam had marked a change in the role of officers. Because of helicopters and better communications, colonels and generals could direct a battle from a distance. Therefore, this was the first war in which the proportion of colonels and generals killed in combat was not equivalent to their proportion in the whole force. In World War II, 77 Army colonels were killed in combat, or one out of every 2,206 deaths, which is in keeping with the ratio of such officers to the entire force. In Vietnam, only 8 colonels were killed, or one for every 3,406 men killed. Not only did this affect the leadership demonstration of these officers, but by directing the battlefield from a remote location, these officers were further removed from any meaningful dialogue with their subordinates.

Moreover, the secretary of defense was Robert McNamara, the former president of the Ford Motor Company. McNamara was an advocate of management by objectives and believed he could reduce the critical aspects of a military campaign to simple numbers. This is an interesting contrast to Dr. Deming's assertion that "the most important things that a manager has to manage cannot be measured." At any rate, McNamara's obsession with numbers gave rise to a massive dependence on meeting numerical goals in the Army. The survey of junior offers found a widespread belief that the Army was corrupt. Most did not plan to reenlist. They believed that the senior officers acted primarily in their self-interest rather than for the good of the Army or the nation.

To quote Walt Ulmer, the senior Army War College faculty member responsible for the study, as he addressed the top Army leadership, "Gentlemen, a scenario that was repeatedly described to us during our interviews for the study includes an ambitious, transitory commander, marginally skilled in the complexities of his duties, engulfed in producing statistical results, fearful of personal failure, too busy to talk with or listen to his subordinates, and determined to submit acceptably optimistic reports which reflect faultless completion of a variety of tasks at the expense of the sweat and frustrations of his subordinates" (Kitfield 1995).

When we first read this, we were shocked at the parallel between this circumstance in the military during wartime and the many similar conditions we see in industry, in which managers pursue numbers and ignore the supposed purpose that the numbers were trying to address. If they can create optimistic reports, by whatever means, they will get bonuses and promotion. Because their tenure in any particular position is brief, they can ignore the long-term consequences of their actions. Likely they will be elsewhere when the long-term consequences are realized.

One of Kitfield's stories is about Jack Galvin, when Galvin was a lieutenant colonel in Vietnam. Near the end of the conflict, Galvin's unit was ordered by his brigade commander to attack an enemy position at dusk. The position was too far afield for support by American artillery. Galvin, knowing that the position was not an important objective and that many of his men would die in the attack, ignored the order. He expected a court martial, but none came. However, in a subsequent fitness report, the colonel stated that Galvin "put consideration for his men above that of mission." The comment was meant to be a condemnation of Galvin's leadership kills. According to Kitfield, Galvin was more proud of that fitness report than of the Distinguished Service Cross or the Silver Star that he won in Vietnam. After Vietnam, Galvin was one of the leaders who helped transform the Army. He rose to the rank of four-star general and ended his career as supreme commander of NATO.

When the results of the morale study were brought to the commander-in-chief, General Westmoreland, it included the following recommendations for action:

- Disseminate the pertinent findings of the study to the officer corps.
- Promote an atmosphere conducive to honest communication between junior and senior officers.
- Add curricula to the Command and General Staff College and other advanced training venues on the subject of interpersonal communications and counseling of subordinates.
- Promulgate an officer's creed to highlight the ethical standards of the officers' corps.
- Take immediate disciplinary action against officers who violate the ethical standards.

Westmoreland's response was to classify the study, which he believed would have very bad political implications. Given that this was his response, it is not surprising that he was presiding over a culture that was dreadfully lacking in open communication and candor. There is good news, however. With the war over and Westmoreland gone, a group of colonels and generals set about to deal with the issues they had uncovered and, as they continued to rise through the upper ranks of the Army, to transform it. The group included Jack Galvin, Barry MacCaffery, and Colin Powell.

Key to the transformation was the establishment of TRADOC, the Training and Doctrine Command. This was perhaps the most critical post in the transformation of the Army. TRADOC was responsible for writing doctrine: how the army would function and how it would fight. But equally important was that it was responsible for developing a training plan so the doctrine would be implemented. The training had to be carried out at every level of the Army. With the issuance of any doctrine, there would be training for recruits at boot camp all the way up to training for generals at the Army War College and the Command and General Staff College. The Army of Desert Storm was the product of this transformation. The initial commander and essentially the founder of TRADOC was General William DePuy. Although few citizens are familiar with his name, Kitfield argues that DePuy was the greatest soldier of his era. Many of his subordinates later rose to important commands in the Army.

An important part of the context for the transformation was the change that modern weaponry dictated on the battlefield. The Army's observation of the Yom Kippur War, in 1973, indicated that modern weapons are so lethal that the next war would likely be over after the first battle. America could not depend on its industrial might to ramp up after an initial attack. We had to decisively win the initial engagement. Based on this perception, DePuy led the creation of the fighting doctrine "air-land battle" that utilized modern communications and transportation and the coordination of forces. This doctrine, executed to perfection because of the Army's effective training, was the basis of the Desert Storm campaign.

Another important facet of the reform was the creation of the after-action review. In both training and wartime, each significant action is reviewed with the objective of finding lessons to be learned. This is an opportunity for subordinates to share information with senior officers that may be critical of the performance of the unit and of these senior officers. Officers who are resistant to this feedback have difficulty with promotion.

Transforming a company is probably less difficult than transforming an army. The process is the same: plan, do, study, act. The U.S. Army began with some critically important data: junior officers felt that the leadership of the Army was working for numbers and promotion not for the good of the Army or the nation or for winning the war. This information is a powerful starting point. According to Kitfield's account, it was a critical source of leverage in the transformation of the Army. At the same time, there is a risk. What is required is to bring information to management that they may not want to hear.

When we brought our original survey findings to the executive committee of a company we had worked with, we had a level of trepidation. We were concerned that as the messengers of bad news, we might be "shot."

In fact, the result was beyond our most optimistic dreams. Years later, the president of the company said that the work we had done, including the survey and the actions it led to, did more to change the culture of the company than anything else done during his tenure. Our work did even more than a multimillion-dollar engagement by a nationally known consulting firm that was specifically targeted at transforming the organization.

TAKING ACTION TO IMPROVE YOUR CAPACITY FOR LEADERSHIP

If this chapter causes you to identify deficiencies in leadership in yourself or in your company, there is still some good news. The ability to lead effectively is a skill that can be learned. It is a difficult skill that may require you to make some changes in the way you have become used to doing your job over many years. Our own experience is that progress can be made, and the results are worth the effort. We now turn to the leadership inventory, a series of questions to help you reflect on the leadership capabilities of you and your organization. This is not a validated survey, and the objective is not to see how your score compares to those of other leaders. Rather, it is a way to see where your strengths lie and discover areas of opportunity for improvement.

LEADERSHIP INVENTORY

Vision

1. I have a clear vision for where our safety process should go.

strongly disagree									strongly agree
1	2	3	4	5	6	7	8	9	10

2. I am enthusiastic about my vision for our safety process.

strongly disagree									strongly agree
1	2	3	4	5	6	7	8	9	10

3. My vision is shared by my colleagues.

strongly disagree									strongly agree
1	2	3	4	5	6	7	8	9	10

Values

Personal

4. I look in the mirror to find who is responsible for our setbacks.

strongly disagree									strongly agree
1	2	3	4	5	6	7	8	9	10

5. I look out the window to find who is responsible for our wins.

strongly disagree									strongly agree
1	2	3	4	5	6	7	8	9	10

6. I am focused on the customers of our safety process.

strongly disagree									strongly agree
1	2	3	4	5	6	7	8	9	10

Company

7. Our company rewards long-term contributions to safety more than short-term contributions.

strongly disagree									strongly agree
1	2	3	4	5	6	7	8	9	10

8. In our company, employees' fear of making mistakes stifles initiative and creativity.

strongly disagree									strongly agree
1	2	3	4	5	6	7	8	9	10

9. Our company continuously strives to improve the safety of every process, practice, and procedure.

strongly disagree									strongly agree
1	2	3	4	5	6	7	8	9	10

10. The leaders of this company are constantly seeking new knowledge.

strongly disagree									strongly agree
1	2	3	4	5	6	7	8	9	10

Communication

11. My subordinates and colleagues have a clear understanding of my vision.

strongly disagree									strongly agree
1	2	3	4	5	6	7	8	9	10

12. My subordinates feel that I hear and understand what they are saying, even when I disagree with them.

strongly disagree									strongly agree
1	2	3	4	5	6	7	8	9	10

13. My colleagues feel that I hear and understand what they are saying, even when I disagree with them.

strongly disagree									strongly agree
1	2	3	4	5	6	7	8	9	10

14. I have a good understanding of the joys and frustrations that front-line employees experience in doing their jobs.

strongly disagree									strongly agree
1	2	3	4	5	6	7	8	9	10

Demonstration

Personal

15. I am willing to risk disagreement in order to properly represent my vision.

strongly disagree — strongly agree

1	2	3	4	5	6	7	8	9	10

16. My actions are always consistent with my statements.

strongly disagree — strongly agree

1	2	3	4	5	6	7	8	9	10

17. My subordinates and colleagues would say that my actions are consistent with my statements.

strongly disagree — strongly agree

1	2	3	4	5	6	7	8	9	10

18. My superiors in the company trust me.

strongly disagree — strongly agree

1	2	3	4	5	6	7	8	9	10

19. My peers in the company trust me.

strongly disagree — strongly agree

1	2	3	4	5	6	7	8	9	10

20. My subordinates in the company trust me.

strongly disagree — strongly agree

1	2	3	4	5	6	7	8	9	10

Company

21. People who do their best for the company's safety process feel adequately recognized by management.

strongly disagree — strongly agree

1	2	3	4	5	6	7	8	9	10

22. Management responds promptly to problems that employees discover.

strongly disagree 1	2	3	4	5	6	7	8	9	strongly agree 10

Impact

Personal

23. I am effective at mobilizing people in our company to confront difficult problems in safety.

strongly disagree 1	2	3	4	5	6	7	8	9	strongly agree 10

24. I have high standards for those who report to me.

strongly disagree 1	2	3	4	5	6	7	8	9	strongly agree 10

25. I persevere and do not bow to obstacles in attempting to achieve our vision.

strongly disagree 1	2	3	4	5	6	7	8	9	strongly agree 10

26. My subordinates see me as a coach rather than as a judge.

strongly disagree 1	2	3	4	5	6	7	8	9	strongly agree 10

Company

27. We have done a good job of removing obstacles to the employee's joy in the workplace.

strongly disagree 1	2	3	4	5	6	7	8	9	strongly agree 10

28. We are aggressively confronting the hard problems that we face as a company.

strongly disagree 1	2	3	4	5	6	7	8	9	strongly agree 10

This inventory can be used to examine your own leadership. It can also be shared with a team or leadership group. We have used it with leadership groups, having the participants return the surveys to us anonymously. We look at two things in the compiled data: (1) Where are the strengths and weaknesses in the group? And (2) where is there a lot of variation in response, meaning that different respondents have very different points of view? Both analyses then provide a basis for discussion of such topics as:

- What are the implications of our weaknesses?
- Why do these weaknesses exist?
- What can be done to improve?
- What are the implications of our strengths?
- How can we best capitalize on them?
- Why are we in disagreement (high variation) on certain issues?
- What is this disagreement telling us about our communication and our ability to assess what is happening with ourselves and with the company?

This chapter represents our own accumulated knowledge and experience as we have attempted to become more effective leaders over a combined 60-plus years in management positions. We put it forward as what we have learned and believe not as any kind of definitive or final answer. We are very clear that leadership skills can be learned. We are equally clear that there are no shortcuts or simple techniques. Leadership is first and foremost about character.

KEY POINTS

- Leadership can be improved through learning.
- Having data that provides insight for future actions can provide powerful support for the exercise of effective leadership.
- Effective leaders focus on the customer, external and internal.
- Leaders are coaches and counsels not judges.

- Leaders use intrinsic motivation by removing obstacles to joy in work.
- Leaders understand variation and do not waste time of subordinates by raising questions about random fluctuations in data.
- Leaders work constantly to improve the system.
- Leaders create trust.
- Leaders forgive mistakes.
- Leaders are humble. They listen to subordinates and are constantly learning.
- Leaders are judged more by what they do than by what they say.
- Leaders mobilize their group to confront the difficult problems and help the group stay calm and focused in the face of adversity.

ENDNOTE

1. A similar listing is found in Deming, 1993, beginning on page 128.

References

Chapter 1

Bassi, L. J., J. Ludwig, D. P. McMurrer, and M. Van Buren. *Profiting from learning: Do firms' investments in education and training pay off?* www.astd.org, 2003.

Carder, B. Kicking the habit [of poor management]. *Quality Progress* 24:87–89, 1991.

Collins, J *Good to great: Why some companies make the leap—and others don't.* New York: HarperBusiness, 2001.

Collins, J C., and J. I. Porras. *Built to last: Successful habits of visionary companies.* New York: Harper Collins, 1994.

Easton, G. S., and S. L. Jerrell. The effects of Total Quality Management on corporate performance: An empirical investigation. *Journal of Business* 71:253–307, 1998.

Feynman, R. P. *What do you care what other people think? Further adventures of a curious character/Richard P. Feynman, as told to Ralph Leighton.* New York: Norton, 1988.

Kohn, A. *No contest: The case against competition.* Boston: Houghton Mifflin, 1986.

Petersen, D. E., and J. Hillkirk. *A better idea: Redefining the way American companies work.* Boston: Houghton Mifflin, 1991.

Whitney, J. O. *The trust factor: Liberating profits and restoring corporate vitality.* New York: McGraw-Hill, 1994.

Chapter 2

Anderson J. W., S. Vichitbandra, W. Qian, and R. J. Kryscio. Long-term weight maintenance after an intensive weight-loss program. *Journal of the American College of Nutrition* 18:620–627, 1999.

Collins, J. C., and J. I. Porras. *Built to last: Successful habits of visionary companies.* New York: Harper Collins, 1994.

Deming, W. E. *The new economics for industry, government, education.* Cambridge: MIT Center for Advanced Engineering Study, 1993.

Festinger, Leon. *A theory of cognitive dissonance.* Stanford: Stanford University Press, 1957.

Marshall, B. *Helicobacter pylori*: 20 years on. *Clinical Medicine* 2:147–152, 2002.

McClamroch, K. Total fertility rate, women's education, and women's work: What are the relationships. *Population and Environment* 18:175–186, 1996.

Petersen, D. Safety management 2000, our strengths and weaknesses. *Professional Safety* 45:16–19, 2000.

Pransky, G., T. Snyder, A. Dembe, and J. Himmelstein. Under-reporting of work-related disorders in the workplace: A case study and review of the literature. *Ergonomics* 42:171–182, 1999.

Rosenbaum, M., K. Vandenborne, R. Goldsmith, J. A. Simoneau, S. Heymsfield, D. R. Joanisse, J. Hirsch, E. Murphy, D. Matthews, K. R. Segal, and R. L. Leibel. Effects of experimental weight perturbation on skeletal muscle work efficiency in human subjects. *American Journal of Physiology-Regulatory Integrative and Comparative Physiology* 285:183–192, 2003.

Senge, P. *The fifth discipline: The art and practice of the learning organization.* New York: Doubleday/Currency, 1990.

Waldrop, M. M. *Complexity: The emerging science at the edge of order and chaos.* New York: Simon and Schuster, 1992.

Wickrama, K. A. S., and F. O. Lorenz. Women's status, fertility decline, and women's health in developing countries: Direct and indirect influences of social status on health. *Rural Sociology* 67:255–277, 2002.

Wing, R. R. Physical activity in the treatment of the adulthood overweight and obesity: Current evidence and research issues. *Medicine and Science in Sports and Exercise* 31:S547–S552 Suppl., 1999.

Chapter 3

Chronbach, L. J., and P. E. Meehl. Construct validity in psychological tests. *Psychological Bulletin* 52:281–302, 1955.

Deming, W. E. *Out of the crisis.* Cambridge, MA: MIT Center for Advanced Engineering Study, 1986.

Gardner, H. Who owns intelligence? *International Journal of Psychology* 35:3–4, 2000.

Jensen, A. R. Psychometric-G as a focus of concerted research effort. *Intelligence* 11:193-198, 1987.

Kranzler J. H., and A. R. Jensen. The nature of psychometric-G-unitary process or a number of independent processes. *Intelligence* 15:397–422, 1991.

Lewis, M. *Moneyball.* New York: W. W. Norton and Company, 2003.

Chapter 4

Heinrich, H. *Industrial Accident Prevention. 4th ed.* New York: McGraw Hill, 1959.

Nelson, L. S. Shewhart control chart—tests for special causes. *Journal of Quality Technology.* V.16 n 4 Oct 1984:237–239, 1986.

Perrow, C. *Normal accidents: Living with high risk technologies.* New York: Basic, 1984.

Petroski, H. *To engineer is human.* New York: Random House, 1982.

Wolf, F., and E. Berniker. *Validating normal accident theory: Chemical accidents, fires and explosions in petroleum refineries.* www.plu.edu/~bernike/NormAcc/Validating%20NAT.doc, 1999.

Chapter 5

Center for Chemical Process Safety (CCPS). *Guidelines for investigating chemical process incidents.* New York: American Institute of Chemical Engineering, 1992.

Center for Chemical Process Safety (CCPS). *Guidelines for investigating chemical process incidents. 2nd ed.* New York: American Institute of Chemical Engineering, 2001.

Chapter 6

Carder, B., and P. T. Ragan. A survey-based system for safety measurement and improvement. *Journal of Safety Research* 34:157–165, 2003.

Carmody, C. J. Testimony before the Illinois General Assembly Committee on Judiciary on Primary Seatbelt Law Enforcement. www.ntsb.gov/speeches/carmody/cc030304.htm, 2003.

CAW-TCA, Canada. Fact sheet: Behaviour-based safety programs. www.caw.ca/whatwedo/health&safety/factsheet/hsfsissueno14.asp, 2003.

Harper, A. C., J. L. Cordery, N. H. de Klerk, P. Sevastos, E. Geelhoed, C. Gunson, L. Robinson, M. Sutherland, D. Osborn, and J. Colquhoun. Curtin industrial safety trial: Managerial behavior and program effectiveness. *Safety Science* 24:173–179, 1996.

Krause, T. R. *Employee-driven systems for safe behavior.* New York: Van Nostrand Reinhold, 1995.

Krause, T. R. *The behavior-based safety process: Managing involvement for an injury free culture*. 2nd ed. New York: Van Nostrand Reinhold, 1997.

Krause, T. R., K. J. Seymour, and K. C. M. Sloat. Long-term evaluation of a behavior-based method for improving safety performance: A meta-analysis of 73 interrupted time-series replications. *Safety Science*, 32:1–18, 1999.

Krause, T. R. *Myths and misconceptions about behavior-based safety: Why conventional wisdom is usually wrong*. www.bstsolutions.com, 2003.

O'Neill, R. It's the hazard, stupid. *Hazards* 7:4–5, 2002.

Ragan, P. T. Behavioral safety: Safety's silver bullet? *Professional Safety* 42:26–31, 1997.

Skinner, B. F. Are theories of learning necessary? *Psychological Review* 57:193–216, 1950.

Walker, C. Behaviour based safety programs, or if it's rat psychology, who is Pied Piper and who are the rats? Canadian Auto Workers, IAPA/WHSC session, "Myth of the careless worker," April, 2003.

Williams, J. H., and E. S. Geller. Behavior-based intervention for occupational safety: Critical impact of social comparison feedback. *Journal of Safety Research* 31:135–142, 2000.

Chapter 7

Bailey, C. W., and D. Petersen. Using safety surveys to assess safety system effectiveness. *Professional Safety* 2:22–26, 1989.

Bailey, C.W. Managerial factors related to safety program effectiveness: An update on the Minnesota perception survey. *Professional Safety* 42:33–35, 1997.

Eisner, H. S., and J. P. Leger. The international safety rating system in South African mining. *Journal of Occupational Accidents* 10:141–160, 1988.

Kuusisto, A. *Safety management systems: Audit tools and the reliability of auditing.* Doctoral dissertation. Technical Research Center of Finland, 2000.

Polk, J. F. Statistical analysis of railroad safety performance, 1977–1982. Final report of contract DTFR 53-82-X-0076. Federal Railroad Administration, 1987.

Pringle, D. R. S., and A. E. Brown. International safety rating system: New Zealand's experience with a successful strategy. *Journal of Occupational Accidents* 12:41–42, 1990.

Ray, P. S., R. G. Batson, W. H. Weems, Q. Wan, G. S. Sorock, S. Matz, and J. Coynam. Impact of maintenance function on plant safety. *Professional Safety* 45:45–49, 2000.

Uusitalo, T., and M. Mattila. Evaluation of industrial safety practices in five industries. In A. Mital, ed. *Advances in industrial ergonomics and safety.* Vol. 1. London: Taylor and Francis. p.353–358, 1989.

Chapter 8

Coyle, I. R., S. D. Sleeman, and N. Adams. Safety climate. *Journal of Safety Research* 26:247–254, 1995.

OSHA. Safety and health program management guidelines; Issuance of voluntary guidelines. *Federal Register* 54:3904–3916, 1989.

Chapter 9

Ackoff, R. *Creating the corporate future.* New York: Wiley, 1981.

Carder, B., and P. T. Ragan. A survey-based system for safety measurement and improvement. *Journal of Safety Research* 34:157–165, 2003.

Kitfield, J. *Prodigal soldiers.* New York: Simon and Schuster, 1995.

O'Toole, M. The relationship between employees' perceptions of safety and its management culture. *Journal of Safety Research* 33, 2002.

Chapter 10

Army War College. *Study on military professionalism.* Carlisle Barracks, PA: U.S. Army War College, 1970.

Collins, J.. *Good to great: Why some companies make the leap—and others don't.* New York: Harper Business, 2001.

Covey, S. R., A. R. Merrill, and R. R. Merrill. *First things first: To live, to love, to learn, to leave a legacy*. New York: Simon and Schuster, 1994.

Deming, W. E. *The new economics for industry, government, education*. Cambridge: MIT Center for Advanced Engineering Study, 1993.

Lindgren, J. Ranking our presidents: How did 78 scholars decide how to rank the presidents from Washington to Clinton? Federalist Society and *Wall Street Journal* survey on presidents. www.opinionjournal.com/hail/rankings.html, 2000.

Index

Note: Page numbers in italics refer to illustrations; a *t* following a page number refers to a table.

A

B

C

F

J

K

L

Q

R

S

W

Y